Humanity at the Crossroads

Technological Progress, Spiritual Evolution, and the Dawn of the Nuclear Age

Garth J. Hallett

Hamilton Books

An Imprint of
Rowman & Littlefield
Lanham • Boulder • New York • Toronto • Plymouth, UK

Copyright © 2015 by Hamilton Books
4501 Forbes Boulevard, Suite 200, Lanham, Maryland 20706
Hamilton Books Acquisitions Department (301) 459-3366

Unit A, Whitacre Mews, 26-34 Stannary Street,
London SE11 4AB, United Kingdom

Library of Congress Control Number: 2015930804
ISBN: 978-0-7618-6561-2 (pbk : alk. paper)—ISBN: 978-0-7618-6562-9 (electronic)

In the days of Louis XIV, France was known as "Le Grand Pays," The Great Country. It was great, not only economically and militarily, but in the strength of its cultural accomplishments. Today, we are "Le Grand Pays," The Great Country. We have sculpted a New World, one which came, in the course of time, to the timely rescue of the Old. Should history move in cycles, should we fear our imminent decline, may we be bolstered by the knowledge that the glory of the sunset, in the cycle of days, is even greater than that of the dawn.

This book is dedicated to The United States, my homeland, a land of bewildering contrasts—of grand achievements and of equally grand mis-steps—but with a civilized heart and a steadfast will to achieve, to en-dure, and to remember, at critical times, the best within itself. May the positive choices we make today lead to a better future tomorrow for those who will soon succeed us, and who now look in our direction for solace.

In the throes of what ultimately may be the direst of human trials, and amidst such harsh and unfamiliar challenges as remain enveloped in our future's obscurity, may we not, out of pain, confusion, or weakness, fearfully or "expediently" abandon the heights of our collective nobility for any temporary or spurious advantage. Should we do so, we are bound to be left, as a result, with only a legacy of ashes and thirty dissatisfying pieces of silver.

When the Way rules the world,
Coach horses fertilize fields;
When the Way does not rule,
War horses breed in the parks.

—Lao Tzu, *The Tao Te Ching*

Contents

Acknowledgments

In presenting *Humanity at the Crossroads* to the reading public, I wish to express my particular gratitude to the following: To my magnificent mother, Gloria Hallett, who had a book in my hands before my legs could walk. She made the right decisions for me, before I could make them myself. To Walter Petry of Fairfield University, undergraduate professor par excellence, whose challenge to his students to transcend mediocrity has been a beneficent "thorn in my side" since college. To Dr. Daniel Buczek, also of Fairfield University, teacher of Russian History, friend, mentor, and, quite simply, the greatest man I have ever known. To UCONN Political Science professor, Richard Hiskes, for his sound and constructive guidance during my graduate school years. To Paramahansa Yogananda and Joseph Goldstein, whose spiritual techniques and teachings informed my outlook, shaped my character, and transported me into adulthood. To Father Joseph Kettner, whose work among the disadvantaged continues to remind me that the world is our larger self. To my tireless editors at the University of Saint Joseph, Antoinette Collins and Jessyka Scoppetta, for casting their distinctive brilliance on the blind spots in my writing. Likewise, to my terrific editor at The University Press of America, Nicolette Amstutz, whose excellence nursed the manuscript to completion. To all those professors and experts at The University of Saint Joseph who read and commented on my work; prominent among them, Shyamala Raman, Director of International Studies and Programs, Caitlin Fitzgerald, Communications Coordinator, Bonnie Solivan, Teaching and Learning Specialist, Kate Stratton, Director of Campus Ministry, Steve Jarett, Department of Business Administration Chair, Jemel Aguilar, Assistant Professor at the Department of Social Work and Latino Community Practice, Judith Perkins, Professor of Classics and Humanities, Sheldon Friedman, Associate Professor of Accounting and Business Management, and University President, Pamela Reid. Their valuable insights contributed to a better result. To all the other nameless giants whose shoulders I continue to stand on. To the University of Saint Joseph, itself, for being an environment in which greatness can flourish. To my wife, for putting up with me (no easy task). To my cat, Oreo, for occasionally refraining from making his own paw-mark revisions to my manuscript. And, last but not least, to my bright and enchanting daughters, Erin and Michelle, members of that emerging generation to whom the future belongs. May the world they inhabit as adults be as filled with hope and blessings as the one my wise and courageous forebears left to me.

Introduction

This book is about the prospective end of nationalistic wars as we have known them, and about the positive and progressive elevation of international competition to more survivable realms of creative human endeavor. It is also about the spiritual evolution of mankind. Its conclusions are based on a perception of trends and events as they are, rather than as we might whimsically prefer them to be. Its envisioned outcome, humanity's collective movement beyond war, would be utterly chimerical, were it not so thoroughly driven by the forces of historical necessity, the exigencies of evolutionary progress, and, perhaps most convincingly, by the conditions for satisfying our more patently selfish ends. The indicated course of this positive movement forward is in the direction of new ways of seeing, thinking, knowing, and interacting that are as yet but dimly perceived—let alone acted on—which are nonetheless critical to our collective development and survival going forward.

This book does not, however, suggest that such a favorable outcome as the abandonment of war is in any way guaranteed, or that anything less than a thoroughly tragic end to all of human life and history will ultimately be realized, perhaps even within our own time; that destiny, our destiny, depends, in the final analysis, on us. What it *does* suggest is that, as human beings, we have been presented with a collective choice between utopia and dystopia, between what President Kennedy, in his inaugural address,[1] contrasted as the wonders and terrors of science. The outcome of that collective choice is one that will shape, not only our own immediate future, but most likely our children's future, and our children's children's future going forward, and perhaps even irrevocably.

Technology, in its ever-expanding role, has so magnified the scope and intensity of human destructiveness that extreme levels of devastation which once "took an army" can now be inflicted by perhaps one individual alone. This ineluctable fact has engendered, as its inevitable concomitant, a pervasive "culture of surveillance," relentlessly implemented and continuously refined, to stave off the next potential crisis, whether of war, of terror, or of ad hoc violence. Yet, without a movement away from war itself, such a capability is likely to be misused, and to ultimately drag us, in the United States, as fully as in other technologically-advanced nations, in the direction of an emerging despotism. Here, the wrong type of international environment, one in which our persistent human flaws are precariously magnified by technology's waxing influence, will inevitably elicit the wrong type of domestic one, with an Orwellian dystopia in the offing. This book sketches the outlines of a far better future and the requirements for its attainment, while emphasizing the unique sense of urgency associated with those critical choices we together must make now.

Still, a question the reader might reasonably ask before tackling the work before him *is*, quite simply, why now? Why, during a protracted period of relative peace and progress, are we focusing, as it would seem unnecessarily and morbidly, on the grisly prospect of war, and particularly on war of a sort that seems nowhere visibly on the horizon? The answer is to be found in the largely unaltered contours of human nature, and in the annals of repetitive violence that constitute our history. There has never been an age on record that has not known its experience of war. While technology has continued to advance exponentially, magnifying and enhancing our raw capabilities, people themselves have remained as obstinately selfish, thoughtless, vindictive, and cruel as they have ever been.

From Benjamin Constant in 1816 to Norman Angell in 1910, there were those, particularly in Europe, who had convinced themselves and others that the civilized world had outgrown its experience of war (Tuchman; Kirsch). As if to emphasize the point, the placid scenes of the French impressionists, composed in the fin de siècle era prior to World War I, denote a continent preoccupied with other than militaristic concerns. Yet such voices of peace spoke too soon, with 1999 marking the close to what had preceded, despite its chronological proximity to our own time, as the most monstrous and belligerent century in all of human history.

Soldiers journeying by train to the front at the outset World War I, whether British, French, or German (among others), were frequently photographed triumphantly waving, as if they were embarking on a quest-like adventure, more like a medieval jousting match than like the unprecedented horrors that actually awaited them along the trench-strewn battle lines and crater-pocked "killing fields" of France. Human nature had retained its traditional visage unmodified, its rough admixture of good and bad; it was technology, functioning as the changed variable, that radically transformed war's anticipated and traditionally more limited outcomes, heightening ominously the intensity and scope of human destructiveness. It is now virtually a truism to suggest that technological growth is accelerating at an ever more rapid clip, making its catastrophic potential, in our current age, that much more considerable.

By way of further explanation, it is also worth noting that times of war are never ideal periods for thoughtful reflection. A considered analysis of war's origins, progression, and larger collective impact, is most likely, during wartime itself, to be interpreted as impertinent dissent, with such dissent itself targeted as a traitorous irrelevance to be violently and mercilessly squelched. That "the first victim of war is the truth" is as true today as it has ever been. It is now, still fairly close to the sleepy dawn of the twenty-first century, as a chronically distressed and war-weary world fitfully catches its breath, that we have been granted the rare and beneficent opportunity to constructively ponder the relationship of our past to our future, and to thoughtfully, rather than haphazardly, craft our direction forward toward a more humane world, a world in which we may hope to subsist as active and thriving participants, rather than as curious bone fragments unearthed from archeological digs undertaken by the next intelligent species to succeed us.

If a driver is falling asleep at the wheel, and appears to be veering off course, no subtlety of argument should be required to convince him, once alerted, to make the necessary adjustments. A sharp nudge, and a finger pointing toward the road ahead, should be enough to prompt him to change direction, to avert impending disaster, and to continue safely on his way.

This book is intended to be one such salutary nudge and a finger pointing in the direction of a more positive future. It is a "wake up call" to a civilization that has, in certain respects—though by no means all—begun to veer dangerously off course. It purports to demonstrate the logical connection between our current actions and their probable future consequences in a way that will hopefully become clear, once it is adequately explained. It identifies some likely future obstacles as they are already hazily coalescing on the horizon of our immediate experience, so that, in being sufficiently aware of them, we can hopefully avoid them, as we proceed on our way to an otherwise promising future.

It has been said that a little knowledge is a dangerous thing. The conflict theories of the 19th Century paved the way for the wars of the 20th, wars that proceeded not only, as all wars do, amidst tragic and heated conflict, but against a theoretical backdrop that seemed to justify such conflict, and its coarse inevitability, intellectually. Darwin observed and catalogued forms of life that were contending for survival within environments of material scarcity, and came up with a theory of evolution, based on conflict, that like Freud's theories of the conflicted mind, were correct as far as they went, but which didn't go far enough. They left unfathomed the grander depths of human thought and striving, amidst the Nietzschean "death of God," and the denial, in the absence of any definitive test-tube proof, of our central spiritual core. Marxism-Leninism, as a conscious approach, applied Darwinism to politics, translating the natural struggle into a conflict between revolutionary classes. Spencerians, somewhat ironically, applied the same Darwinian framework to the capitalist economy itself, utilizing it as a rationale for ongoing social inequality. Even Hitler's grotesque theories of race-based struggle were a twisted and particularly unscientific offshoot sprung from the same theoretical tree.

The consequential question arising from the application—or misapplication—of such ideas and theories of conflict, as they relate specifically to the course of our future planetary development, is not whether humankind was spawned through evolution or creation, as a matter less consequential than speculative, but whether we are intended by nature to love our fellow man or to bury him. In the nuclear age, this question becomes more pointed, and if the former should indeed be true, then we need to look deeper than the surface observations of 19th Century empiricism for a fuller understanding of our quintessential human nature, and for a more enduring solution to the problem of weapons and war.

In life, as in math, a negative (in this case the darkness of our instinctive nature) multiplied by a greater positive (scientific and technological advancement) results only in a greater negative (in unnaturally destructive wars or, alternatively, as this book describes, in the compensatory rise of dystopian societies). To know is to grow, making knowledge good in its independent nature, with scientific and technological progress, part and parcel of our broader and more expansive development. Yet knowledge in service to ignorance, as this book argues, only intensifies the impact of that ignorance, and the increasing threat that it poses along the uncertain road ahead. This is the pivotal problem and central evolutionary challenge facing humankind in the modern age.

Technology, in the modern world, has served as a great magnifier and illuminator. Where we have been destructive, it has heightened our capacity for destruction, and the consequences of collective violence. Where we are creative, it has enhanced and rarified the creative output. In a fundamentally unified world, it has accentuated the underlying unity that has always been there. It has, in the process, served as the primary catalyst for, and concentrated accelerant of, contemporary change, highlighting what we need to embrace, and blue-penciling what we urgently need to abandon. In the process of our hopefully "cleaning up our act," it is prodding us to evolve in the direction indicated by the great religious traditions of the world, traditions whose truths in the modern age appear not only profound but prophetic, in being more impactful than ever before. In the modern era, failure to abandon our atavistic evolutionary baggage will either cause the requirements of national security—and the ensuing curtailment of liberty—to leaven unnaturally or, by enhancing the destructive potential of terrorism and war, lead inevitably to catastrophic losses of life, up to and conceivably including our own extinction.

Stopgap measures, such as punitive restraints on nuclear proliferation, and efforts to bolster and safeguard our independent security, while undeniably crucial and necessary, are, as this book suggests, ultimately doomed to failure if relied upon exclusively. For they focus only on the positive side of the aforementioned equation, the side that can only increase exponentially with time. In fact, any attempt to manipulate or to curtail the natural technological progress of any region, and particularly of polities already inimical to us, if pursued indelicately, can add at once to the negative side of the equation and to its greater mathematical result. This would inevitably be true, as any constraints prompted by ends not their own are very likely, by such nations, to be resented, if not rejected "out of hand."

As this book explains, we are moving through the spiritual and material processes of evolutionary expansion toward the peace of an integrative world. This is not, however, because, (and as demonstrated by the circumstances surrounding the Cuban Missile Crisis), that is where we originally intended to go, but because that is where the changed requirements of our new technological environment have compelled us to be. The security that was previously realized through war, can now be attained at the highest international level, and by the most developed nations, only through the principles and methods of peace. As this work argues, where each seeks security at the expense of the rest, all together are insecure. Where each seeks peace as an isolated betterment, the ultimate end result is war.

Humanity at the Crossroads further addresses the flawed yet characteristic way we tend to view the world and our separate advantage in it. It demonstrates how "misperception leads to misconception leads to war." Here, dire results often proceed from an inability or unwillingness, in the very first instance, to see life as it is, or to define our interests at an appropriately elevated level. *Humanity at the Crossroads* charts the mechanism by which such misperception and misunderstanding occur. It further shows how problems of perception have lulled us, for the moment, into believing that the problem of all-out war is largely a legacy issue, leaving us even more vulnerable, in the future, to the probable onset of its brutal and cataclysmic unfolding.

Albert Einstein reputedly asserted that a problem cannot be resolved at the same level of thinking (or of consciousness) at which it was created. As this book describes, there is a dynamic associated with wars that parallels the one associated with our peaceful development. If nation A wants peace, B wants war, and each is determined in its resolve, then A and B will go to war. A more positive result might well be obtained at an individual level, where the consequences of capitulation are subdued, and the context of action and behavior, more constrained; but, at the national level, this truth is so axiomatic, it can be identified, for the purposes of this introduction as "Rule 1." As demonstrated by the premier example of Hitler's war in Europe, such heated events unfold at a level, and at a point in time, at which the problem of war *cannot* be resolved. One must look to a broader context, and to an earlier time (not just to Munich and appeasement), to such pertinent factors as the punitive nature of German reparations and the infamous War Guilt Clause to discover a more complete and satisfactory explanation as to why World War II occurred. We may thus hope to learn the lessons of that war, so as to forgo similar conflicts in the future.

Rule 1 is closely tied to what this book describes as "the principle of the moving inchworm." This principle states that we can only advance so far beyond the level of our least advantaged members without being constrained in our progress by their backwardness. If we leave the globe's "lesser nations" too far behind for too long to fester in their separate discontent, they will invoke Rule 1 as the most obvious, and perhaps only means of garnering the world's attention, in the hope of satisfying their separate collective aspirations. In the process, the more advanced nation or nations who have themselves evolved beyond the need for war, and the recourse to violent means, will be dragged back down involuntarily to that same warlike level, diverted from the natural course of their greater and more peaceful development by the need to cope with the impact of sporadic regional conflagrations. A number of pertinent historical examples are cited to demonstrate this point. This book argues further, in a "bigger they are, harder they fall" scenario, that the farther ahead the more advanced nations progress, the more disastrous is likely to be any prospective "fall from grace" in the abrupt reversion to the violent methods of the past.

Tied to these two interrelated principles is the startling dynamic of planetary evolution, already briefly alluded to, by which ongoing advances in technology, including the growing sophistication of military arms itself, is propelling humanity at large in the direction of an ever-more-integrative world. As time goes on, technology becomes more enabling. As technology becomes more enabling, the international ties that bind, particularly in the realm of economics, become firmer and more numerous, while the invocation of Rule 1 becomes correspondingly more problematic. This circumstance presents us, in the final analysis, with a collective choice between a peaceful future realized on a roughly equitable basis worldwide, or no future at all, at least not one worth living, for we will not be able to inhabit that future freely, if we have not at first prepared for it compassionately and wisely.

On a more positive note, as this book argues, we continue with the trend of weapons and war, not because the human species is incorrigibly warlike, but because the root causes of war have not themselves been adequately addressed. This makes war, less of an ineluctable cause for unhappiness in the modern world, than a highly unfavorable effect of our inability to think and respond appropriately, of our insensitivity, ignorance, and neglect. *Humanity at the Crossroads* examines this soft-pedaled "negative" side of conflict's mathematical equation, one in which the outlines of a positive, practical, and lasting end to the age-old menace of full-scale military conflict can at last be discerned.

The book's second half identifies some specific obstacles to progress in the hope that we may safely overcome them. It is its author's fervent wish that this book, as a whole, will contribute meaningfully to the emergence of a more positive, free, and peaceful world, and to a more thoughtful and compassionate understanding of ourselves, and of the larger needs of Life.

NOTE

1. The complete reference, from John F. Kennedy's January 20th, 1961 Inaugural Address, is "Let both sides seek to invoke the wonders of science instead of its terrors. Together let us explore the stars, conquer the deserts, eradicate disease, tap the ocean depths, and encourage the arts and commerce."

I

Old Endings and New Beginnings

Chapter One

The Turning Point

With the dawn of the thermonuclear age,[1] and the ensuing deployment of strategic nuclear arms by the opposing sides, humanity reached a critical threshold, a turning point in our collective development at which efforts to ensure our national survival threatened us with an unnatural extinction.

Conditions had even begun to suggest that we were becoming biologically obsolete within the artificial, technological environment we had created for our benefit. Technology had so distorted the cause and effect relationship associated with the conditions of nature, that instincts which had promoted our survival in the natural state, now threatened us with extinction in the civilized state. Aggressive impulses, which aided us in our battle against natural enemies, now served as the catalyst for mass-annihilation in a substantially altered world in which Man's most notable enemy was Man himself.

In a Taoist sense,[2] the development of nuclear arms marked that point of paradox within the context of human history at which one half of the duality which undergirds the phenomenal universe is transposed into its opposite. Nuclear weapons were so destructive as to be effectively useless as weapons of war. They left the strong, who possessed them, less secure overall, and the weak, who lacked them, with unique bargaining advantages over the strong. The most nightmarish weapons of war, they inaugurated an era of peace we would hardly have enjoyed in their absence.

Nuclear weapons, in their strategic variant, have threatened to sacrifice the human species as a means of defending the more partial and exclusive interests of particular national groups. Their use, unlike that of stones and spears, would only defeat their purpose, for with them, the ends of conflict are achieved only to the extent that conflict is avoided. While their value is thus strictly in their non-use, it is nonetheless dependent on our continued willingness to use them if necessary. Such a unique and convoluted method of defense runs counter to what our natural and historical conditioning have led us to regard as the appropriate means and anticipated results of conflict. It leaves much to be feared, despite the prevailing climate of peace, in how little our natural experience has prepared us to restrain our violent instincts amidst the unprecedented over-effectiveness of our capabilities. As noted by Kosta Tsipis, physicist author of *Arsenal: Weapons in the Nuclear Age:*

> Since the time that people stopped being gatherers and hunters and organized themselves in the first villages and communities, it was evident that by killing an enemy who threatened one's welfare and life, one's own survival was enhanced. But now, with nuclear weapons and unstoppable delivery systems that can completely devastate a country, this fundamental assumption is no longer valid. No matter who starts a nuclear war, no matter who strikes first, if one nuclear country tries to destroy the population of a second nuclear country, the second country can assuredly destroy the population of the first. So now killing an enemy does not assure one's own survival, but rather one's own annihilation (253-4).

That we can deny life, yet not, in the process, deny but enhance our individual positions, that we can obliterate ambient nature and still remain standing, is a perception which is valid (if truly at all) only within the context of naturally-limited means. Amidst the substantially changed context of nuclear capabilities, the "other" we move to destroy becomes one in the moment of conflict with the "self" we seek to preserve. A pragmatic brotherhood of man thus emerges as an outgrowth of the individual will to live, in a conflict environment in which "no conflict" becomes the most appropriate response. This is the most dramatic of all paradoxes associated with the existence and influence of nuclear arms. It, among others, will now be further discussed.

NOTES

1. Development of the hydrogen bomb was authorized on 31 Jan. 1950; the U.S successfully test exploded a hydrogen (thermonuclear) device in the Marshall Islands on 24 March 1954 (Carruth).

2. The Tao Te Ching, a work attributed to the Chinese mystic Lao Tzu, is replete with examples of how a particular state is attained by means of its opposite, the Yang by way of the Yin. As noted by R.B. Blakney in his introduction to The Way of Life (Tao Te Ching): "The Yin-Yang experts regarded the interaction of these cognates as the explanation for all change in the universe" (Tzu 28). Some "Taoist paradoxes" from The Tao Te Ching are "the wise man chooses to be last, and so becomes the first of all / Denying self, he too is saved" and "the stiffest tree is readiest for the axe" (66, 152).

Chapter Two

The Hinge of Paradox

The development and deployment of nuclear weapons, during what has historically been termed the Cold War (something of an oxymoron in itself), was characterized by a number of complex and interlacing paradoxes. Giving voice to what is perhaps key among them, military strategist Bernard Brodie once observed, "Thus far the chief purpose of our military establishment has been to win wars. From now on its chief purpose must be to avert them" (76). He also claimed, in another mind-bending, koan-like assertion, that, "War is unthinkable but not impossible, and therefore we must think about it" (qtd. in Kaplan 34). As Kaplan summarizes Brodie's resulting viewpoint, "To be effective, a defense against an atomic attack must be *completely* successful, and the annals of military history provided no encouragement for believing such perfection possible" (Kaplan 26).

The intricate nature of nuclear defense was further indicated in numerous other comments, uttered in frustration or perplexity during the course of the Cold War. This included one by John Foster Dulles, who stated near the end of Eisenhower's first Presidential term, "Of course we were brought to the verge of war. The ability to get to the verge without getting into war is the necessary art. If you cannot master it, you inevitably get into war" (qtd. in Newhouse 96).

The condition that, with thermonuclear arms, in particular, the ends of conflict are achieved only insofar as conflict is avoided, that their value is strictly in their non-use, yet dependent on our willingness to use them as necessary, was neatly encapsulated in one of William Kaufmann's Cold War statements on the essential role of deterrence. As Fred Kaplan explains from *The Wizards of Armageddon*:

A policy of deterrence, Kaufmann recognized, inevitably carried a potentially costly risk—"that despite our best efforts, the antagonist will challenge us to make good on our threat. If we do so, we will have to accept the consequences of executing our threatened action. If we back down and let the challenge go unheeded, we will suffer losses of prestige, we will decrease our capacity for instituting effective deterrence policies in the future, and we will encourage the opponent to take further actions of a detrimental character." If the threat is massive retaliation and if deterrence for some reason fails, then the only way to avoid perilous humiliation is to go ahead and drop atom bombs: yet the Soviet Union can also massively retaliate in return. "In other words," Kaufmann concluded, "we must face the fact that, if we are challenged to fulfill the threat of massive retaliation, we will be likely to suffer costs as great as those we inflict." And that is unacceptable (190).

Another paradox, and rather odd perceptual issue that accompanied the initial wartime use of the atomic bomb at Hiroshima, was that the impact of the weapon was so destructive as to be inadequately perceived at the time. As described by George Marshall, "What we did not take into account . . . was that the destruction would be so complete that it would be an appreciable time before the actual facts of the case would get to Tokyo" (qtd. in Newhouse 50-51).

A further notable twist, one emerging this time from a deliberate U.S. policy, stemmed from President Eisenhower's preoccupation with maintaining a balanced budget. This internal priority manifested externally as a campaign to deploy nuclear arms "on the cheap" as part of a concerted effort to counterbalance Russia's preponderance of conventional forces in Europe. This was an all-too-convenient solution to a pressing problem, hungrily embraced at the time, without any systematic awareness as to how such weapons would be usefully employed under actual wartime conditions. As described by John Newhouse in *War and Peace in the Nuclear Age*, "Only by deploying nuclear weapons, it seemed, could the United States afford to meet its commitments, especially in Europe. But actually using the weapons except in the event of an all-out war in Europe was for Eisenhower close to unthinkable" (91).

The troubling implications of this half-solution promptly hit home during the Cuban Missile Crisis, when it was feared by ExComm planners that a U.S. attack on Cuba would be parried by a Soviet invasion of West Berlin. The predictable Russian response could then be met only through the preplanned use of nuclear arms. Such an exaggerated, yet seemingly given rejoinder, would, in turn, have caused the U.S to lurch ever closer to an all-out nuclear exchange. Still, despite lingering doubts as to its genuine applicability, the deployment of nuclear arms in lieu of conventional forces, otherwise known as the doctrine of "existential deterrence," remained in effect through the end of the Cold War (Newhouse 90-91).

Another noteworthy paradox was associated with MAD, with Defense Secretary McNamara's pet doctrine of Mutually Assured Destruction. According to MAD, one side is virtually guaranteed to obliterate the other regardless of who strikes first; this restrains even the most determined adversary from undertaking any military action likely to precipitate a full-scale war. The paradox here is that a numerical superiority in strategic arms does not make the side possessing it more, but actually less, secure; it does so by casting into doubt the rough assurance of balance, while adding a merely gratuitous weight to an already imponderably destructive equation. In fact, any perceived superiority, whether based on Multiple Independent Reentry Vehicles (MIRVs), Anti-Ballistic Missiles (ABMs), or the Strategic Defense initiative (a purely fanciful alternative to MAD at the time it was initially conceived) would only threaten, as its most immediate and predictable consequence, a toppling of "the wobbly apple cart" on which the precarious peace depends.

That the standard rules of textbook math did not themselves "add up," as applied to nuclear arms, was comparably expressed in a point made by a frustrated Henry Kissinger (Newhouse 245) upon being presented with the prospect that the Pentagon had blocked yet another Strategic Arms Limitation Talks (SALT) agreement. He exclaimed at the time, and in profound exasperation, "What in the name of God, is strategic superiority. . . . What do you do with it?" He noted, in a similar vein, in a speech before the American Legion, that "we, as well as the Soviet Union, must start from the premise that in the nuclear era an increase in certain categories of military power does not necessarily represent an increase of usable political strength. . . . The overwhelming destructiveness of nuclear weapons makes it difficult to relate their use to specific political objectives and may indeed generate new political problems" (qtd. in Newhouse 252).

The ultimate and concluding paradox of the Cold War, that its adversaries were destined to become one in ruin if not one in peace, was aptly analogized by J. Robert Oppenheimer (in a metaphor borrowed from Vannevar Bush). It compared the two main Cold War adversaries to "two scorpions in a bottle, each capable of killing the other, but only at the risk of his own life" (qtd. in Newhouse 92). Policy by other means, in the traditional Clausewitzian[1] sense, the new world of nuclear strategy certainly was not.

A concomitant effect of nuclear weapons, in their Cold War deployment, that the weak who lacked them possessed bargaining advantages over the strong, manifested not only as an occasional anomaly, but as an implied tenet of routine Cold War diplomacy. Here the paramount need to avert nuclear catastrophe sparked a determination on the part of both superpowers to avoid any direct confrontation at all likely to precipitate it. Their all-too-destructive arsenals served to cancel each other out, causing the sledgehammer approach of war to give way, in much of its traditional sphere, to the subtler methods of diplomacy. Lower-tier conflicts rose to an exaggerated level of prominence, while second-level powers assumed an artificially enhanced level of influence. Second World nations in key strategic zones, rather than being unceremoniously conquered, came to be more seductively wooed than Anna Karenina. Discerning the superpowers' predicament and capitalizing on it, many "lesser" nations came to master the art of playing one colossus off against the other, eventually surrendering their loyalty, like a tentative vow, to the most promising suitor. As described by John Stoessinger, "If the United States demanded unacceptable political conditions, the recipient was now in a position to say: 'If you don't give me aid without strings, I know someone who will'" (202). He cites Egypt's construction of the Aswan High Dam, as an historical case in point.

In 1956, as Stoessinger observed, the U.S. offered to finance construction of the Aswan Dam only shortly before Nasser, the Egyptian leader, bought arms from Czechoslovakia and recognized the legitimacy of Communist China. On learning of these developments, Secretary of State John Foster Dulles retracted the American offer. Nasser then requested, and promptly received, alternative aid from the Soviets. This took the form of both Soviet funds and the assistance of Soviet engineers. Its alliance with the Russians cemented at least for the time, Egypt in the 60s became one of the five largest recipients of Soviet aid (Stoessinger 210).

Soviet fortunes in the Middle East were nonetheless fated to decline. In 1967, a good part of the Russian investment was lost as almost three billion dollars in military equipment granted in aid to the Egyptians was either captured or destroyed by the Israelis (Stoessinger 211). A decade later, as described by Alvin Rubinstein:

> Sadat (*Egypt's new leader*) dashed the Kremlin's expectations of a key role in the (*Camp David*) negotiations by resuming diplomatic relations with the United States and plumping his eggs in Washington's basket, leaving the USSR empty-handed, angry, and odd-man-out in the peace process whose deliberations under a U.S.- imprimatur were profoundly to alter the political environment of the Arab-Israeli sector of the Arab East (774).

For the superpowers, conventional foreign policy in the nuclear age took on a correspondingly unconventional aspect.

This peculiar phenomenon of inverted influence was even better exemplified and over a longer term by Russia's symbiotic relationship to Cuba. Here Castro, in an adroit yet precarious gambit, cashed in on the Soviet's largesse, while drawing himself and his regime squarely and hazardously into the nexus of America's nuclear crosshairs. With all the lavish attention paid to him over the years, by both his friends and his enemies, and amidst repeated efforts by the latter to have him forcibly removed, the clearest result of this strategic partnership—amidst the depletion of the Soviet coffers and the enhancement of his northern neighbor's annoyance—was Castro's political and physical survival.

NOTE

1. "The Prussian military thinker Carl von Clausewitz is widely acknowledged as the most important of the classical strategic thinkers. Even though he's been dead for nearly two centuries, he remains a powerful living influence: the most frequently cited, the most controversial, and in many respects the most modern of strategic theorists" (Bassford).

Chapter Three

Beyond the Turning Point

With nuclear arms, as but the first and most prominent in a series of weapons of mass destruction, mankind has entered what may be termed a "species trial." The higher we climb the ladder of knowledge, the greater the damage done should we slip and fall back into the domain of instinct. The more developed the level of science and technology, the more hideous the effects of their abuse.

If we prove ourselves worthy custodians of the vast knowledge attained, we will advance to heights of achievement as yet undreamed; if we do not, we will condemn ourselves collectively to extinction. Here, the nuclear weapon is itself the greatest—though not the only—symbol to date of what failure in this contemporary trial would mean for mankind; nor does the fact that The Cold War is over, fundamentally alter the overall risk for the future.

Suppose you were invited to play a game in which your initial chances of losing were as slight as one in a million. If you won, and continued to win, you would gradually increase your earnings. But if you lost—only once— you would lose everything, and be left empty-handed. Nor would you likely have the chance to ever play again, at least not at that equivalent level. Should you decide to just quit and leave the table, you would run the alternative risk that your winnings may be stolen in your long walk home through the bad neighborhood.

Nations in the nuclear age are confronted with a similar dilemma. So long as we base our strategic defense on the threatened use of nuclear arms, we remain relatively impervious to attack. Particularly now, with the progressive development of a new, more cooperative, world order, the threat of nuclear war, apart from the ad hoc threat of terrorism, seems blissfully remote. In addition, were it not for the existence of nuclear weapons, we would run the risk of suffering grave and repeated losses in more routine conventional wars. Such wars, the equivalent of the long walk home, while consequently more likely to occur, would be admittedly far less damaging than an all-out nuclear exchange.

Yet, even should we now, in acknowledging the long term threat posed by nuclear arms, try to revert, at the strategic level, to more conventional modes of defense, we would be effectively unable to do so. Nuclear weapons have been around since the 1940s. They shall continue to exist as an accessible option for nations that already have them, and will invariably be sought and acquired by others who currently do not. The fundamental reality that these weapons represent of a technology grown threateningly awry unquestionably will not change. Nor are any follow-up equivalents—anti-matter bombs for example—likely to prove any less destructive. The technological marker has irrevocably been passed, and any furtive longing to return to a bygone world in which nuclear arms are absent can never in truth be realized. This would inevitably pertain, at least so long as such weapons continue to be deployed, the motives behind their deployment remain, and war itself retains its status as an acceptable instrument of policy.

In playing the game, thus described, the odds of coming out ahead seem great. Yet the longer we play, the more we run the risk that we *will* eventually lose, and that such an utterly unacceptable one-in-a-million loss will occur. This makes the early obscurity of such an ultimately catastrophic result an additional hazard associated with nuclear arms. In the ebb and flow of politics, the current favorable climate may well, at some point, deteriorate, while we have, as a species, a limitless future in which to make good on both our threats and our promises. The road to that one-in-a-million chance is one we cannot afford to take. Whether or not future societies choose to base their strategies of defense on the actual or threatened use of nuclear arms, the fact that such arms exist—a truth that cannot be erased—has altered forever the nature of conflict, and the human role within it. A threshold of knowledge and capability has been reached at which the rules of life and death have changed.

Increasingly, our ability or inability to adapt to the new ethical demands of the modern world which we ourselves have forged, and of which nuclear arms are a prominent and particularly troubling reminder, will be the primary factor in determining our future survival or extinction. As noted by Sigmund Freud in the concluding paragraph of *Civilization and Its Discontents*:

> The fateful question for the human species seems to me to be whether and to what extent their cultural development will succeed in mastering the disturbance of their communal life by the human instinct of aggression and self-destruction. It may be that in this respect precisely the present time deserves a special interest. Men have gained control over the forces of nature to such an extent that with their help they would have no difficulty in exterminating one another to the last man. They know this, and hence comes a large part of their current unrest, their unhappiness and their mood of anxiety (92).

Chapter Four

Out of Division and Darkness

In the past, full-scale conflict typically required the prior mustering of belligerent populations and the shaping of conscious attitudes in preparation for war. The choice to go to war among certain Native American tribes would, for example, be attended by elaborate rituals (Williams 167-8). Full-scale war in the current age of nuclear missiles, smart bombs, and rapid deployment threatens increasingly to become less of a collective endeavor. It has become more of a mechanized routine, no longer requiring the participation or consent of the majority of those involved. While media technology, as during the Vietnam War, has often served to heighten popular awareness, and to focus collective outrage against unpopular deployments, such responses are now more likely to occur only after the battle is joined.

During the Cold War, U.S.-Soviet rivalry threatened a brand of total conflict to be fought, at least in its initial stage, almost exclusively by machines. Levels of destruction, which in the past could only have been the result of continuous battle, would, in a nuclear exchange, take place within only hours or minutes. In the aftermath of such a war, there would be no popular folklore of conflict, no celebrated acts of individual heroism, only the automated functioning of a belligerent intent which left our humanity and our needs behind.

Human instincts developed at a time when continuity could be expected in the character of our living environment. Responses appropriate to our defense in the past seemed likely to serve our needs in the future. The enemy that attacked with rocks and spears one year would invariably attack with rocks and spears the next. With the advent of industrial technology, that predictable relationship changed. Dramatic shifts in the character of our living environment, brought forth through technology's influence, make it difficult for us to perceive, interpret, and appropriately respond to the modulating

21

threats which face us, including the threat that war itself may pose to human survival.

While it had been maintained by Spencerian Darwinists that primitive warfare promoted "the survival of the fittest," modern warfare, to an ever-increasing degree, has not. The more technology has progressed, the less our evolution has been facilitated—if it ever was in any way for human beings specifically—by a dynamic of physical conflict and violence. The famous colt revolver of the Old West was called "the equalizer" because it made any man in terms of strength the equal of any other, with technology attenuating the role of certain natural advantages (Rosa and May 48). In World War I, the process of natural selection seemed almost to work in reverse. The fittest, who were the first to be drafted, were merely the first to die under conditions where any and all natural advantages were all but wholly offset by the over-whelming destructiveness of mechanized weaponry. At the Battle of Verdun alone, a total of approximately three-quarters of a million men were butch-ered. As H. Stuart Hughes describes it:

> The Germans, intending to bleed their enemies white, had done the same to themselves. The Somme had added half a million German casualties (*to Ver-dun*), 410,000 British, 190,000 French. Human imagination could not grasp such a fearful toll. And for nothing: at the end of the year, the front ran only a few miles from where it had stood at the beginning (63).

Modern warfare conditions, different from those in nature to which our combative instincts pertained, made the European nations' primal longing for greatness the catalyst by which they, both "winners" and "losers," were artificially reduced to second-rate powers in the aftermath of two bloody world wars. If the violently-manifested quest for prominence could be viewed as a collective extension of the individual will to live, then what Nature urged these nations to do for the sake of their individual survival had much the opposite result.

With strategic nuclear arms, the inversion of the natural causality of vio-lent action to its anticipated result reaches its culmination. With their use, not only are natural advantages rendered meaningless, but the very distinction between winning and losing, and from an evolutionary standpoint, the utility of war itself. Whatever process of natural selection may arguably have been served by war in the past, can no longer be served by that means.

It may thus be optimistically ventured, with human adaptability rising to meet the evolutionary challenge, that the warlike phase of our historical trek through time may be gradually winding down, as emphasis shifts from the physical to the ideational, away from violence and toward the development of those higher qualities specifically associated with the rise of civilization. The dawn of the nuclear age, in marking that specific point at which all-out war among first-level powers becomes patently absurd, may, as such, be regarded as a watershed in humanity's evolutionary experience.

As Lipton and Bhaerman comparably observe in their book, *Spontaneous Evolution*:

> The seemingly insurmountable crises that currently challenge our existence can be taken as an obvious portent of civilization's imminent demise. However, below the turmoil evident on the surface, there is an even deeper, more profound reason as to why our civilization is ending. The core beliefs upon which we have built our world are leading us to our own extinction—that's the bad news.
>
> The good news is that new-edge science has drastically revised our current paradigm's core beliefs. By definition, revisions of paradigmatic beliefs inevitably provoke a profound transformation of civilization as its population assimilates the newer, more life-sustaining awareness (205).

This basic transition can be described in Hegelian terms. Our violent primitive nature is the thesis, the technologies of total war, their antithesis. The resulting synthesis is an altered relationship of man to man, and of man to environment, in which conflict is not manifested violently, and with an aim to dominate, but unfolds within a larger cooperative framework of economic, technological, and social change, with the aim of collectively creating a better life. This has occurred in an altered environment in which full-scale violence among the world's major powers can no longer rationally serve the ends of Nature.

We have seen evidence of this change in the end of the Cold War and in the collapse of Soviet Communism. In the Soviet Union, Joseph Stalin based his totalitarian rule on the notion of total war with the West, on the violent incompatibility of competing social systems. As the Soviets achieved nuclear parity with the U.S., this view was progressively modified, giving way, at least in theory, to "peaceful coexistence," to "détente." Still, as in earlier phases, the successes of Soviet Communism remained largely military; they notably failed to extend to the progressively more significant domains of sustained economic development and genuine social progress. Under Gorbachev and later under Yeltsin, the moribund Soviet Union, and then the new Russia, abandoned the outdated modality of gridlocked military rivalry to participate in the emerging network of worldwide economic competition. The Soviet system, rooted as it was in military methods and achievements, became itself obsolete.

The startling relinquishment by the Soviet Union of its East European satellites is notable in this regard as well. In the past, a nation's status was determined overwhelmingly by its "physical" attributes, by territory typically gained through war. For the Soviet Union, the acquisition of an empire in Eastern Europe was, in particular, prompted by largely military motives.

These territories served, not only to enhance the status of the Soviet Union as a military superpower, but were to provide a buffer zone in the event of an invasion. The more relations with the West came to be defined in other than traditional terms, the more nuclear parity rendered Russia effectively invulnerable, the less ownership of these territories meant. A point was reached where their stewardship became more of a burden than a blessing.

As a comprehensive explanation for the collapse of communism, this description is, of course, inadequate. It fails to consider, for one, how the Soviet Union's own internal mismanagement and lagging infrastructure may have made administration of a larger empire unfeasible. Yet it does suggest how a progressive global realignment contributed ultimately to a basic transformation in a major power's collective outlook.

Mikhail Gorbachev, under whose auspices the above sea change was initiated, was a managerial technocrat in the modern Western mode. Boris Yeltsin, though perhaps less Western in manner, continued with modernizing reforms. Yet, the persistent ache of scarcity, as continuing to manifest in critical areas of development (a Malthusian factor of key importance which will later be further considered) may still threaten to degrade Russia—and with it the world—to an atavistic level of confrontation, belligerence, division, and violence which can newly threaten global stability and peace.

We have witnessed a hint of this greater and more ominous possibility in Russia's more limited forays into Crimea and the Ukraine. Such actions, as coarse and belligerent as they may appear—and actually be—cannot be accurately viewed as any attempt to reestablish a buffer zone, along traditional lines and as justified by earlier concerns, but rather as an attempt to recoup territories affiliated with the Russian homeland culturally, historically, economically, and linguistically. These are territories that can realistically be regarded as having been excessively "bartered off" during the "end-of-Cold-War" clearance sale. The fact that the United States and greater world community responded to this unwelcome incursion primarily through the instrumentality of economic sanctions speaks volumes about the underlying importance of such global economic ties within the substantially changed context of a now more unified world.

During the Cold War itself, the frenetic drive of each adversary to keep pace with the other was typically interpreted as an effort to achieve supremacy; hence the race for arms was driven still faster at even greater expense to achieve results that only served to maintain the relative balance. And, although we could visualize the end of the dark road we were travelling, we felt compelled to take it anyway. More bombs were produced in a world where more people starved, and an awareness gradually dawned that the process itself was accelerating out of control, moving inexorably along the rails of its own internal logic, toward the realization of ends opposed to the needs of mankind, toward the spawn of Hollywood's *Terminator*, or *Colossus*.

The mechanism that drives us, as if deterministically, to develop ultra-destructive technologies such as "the bomb" is unbridled competition among members of our own species. Competition can be beneficial, if pursued in a cooperative context. It spurs the individual to growth, and the nation to develop the full intellectual capital of its citizenry. Yet, it can be decidedly detrimental to our planetary survival, if incongruently expressed through the outmoded means of war, violence, and unrestrained antipathy. It can—and must—be directed into channels more suited to our current stage of development, if our technology is to elevate us and not to destroy us. The history of military rivalry between the former Cold War superpowers emphasizes how violent modes of response in the current age of advanced technology can no longer serve our collective needs.

In the modern world, national status and progress is predicated, not so much on raw military strength, as on the ability to compete commercially and technologically. The way this works has been ideally exemplified by Japan which, though a small entity territorially, has, since World War II, made rapid strides in the concentric realms of economic progress and commercial development. A stage was reached in the 80s where, under the changed rules of the game, Japan was able to buy up with relative ease swathes of prime real estate in the very country (i.e., our own) that it had been previously unable to restrain through force of arms.

In the postwar era, the United States has dedicated a substantial percentage of its material resources and engineering talent to producing more and better weapons. Japan, in renouncing war entirely, and in hitching its defensive star to the wagon of protection which America provides, has been able to allocate a far greater percentage of its material and human resources to producing high-tech consumer goods. While we were gloating over our pyrrhic victory against a petty tyrant in the first Gulf War, Mazdas and Toyotas were replacing Fords and Chevys in the garages of American consumers.

Japan and Germany are excellent—though by no means singular—examples of how such a lofty goal as the end of war is made feasible, not just ultimately through a progressive evolution in personal, societal, and international norms, though such a broader advance is critical, but through more routine developments in the largely selfish domain of international economics. These developments have been comparably abetted by technological progress. Advances in telecommunications, in particular, have made a worldwide electronic network of overlapping financial markets an integral part of modern life. These markets have seamlessly interlinked the destinies of nations at no less than light speed, including the futures of those polities otherwise prone to be mutually adversarial.

As is well worth reemphasizing, the modern method of securing wealth, nationally as much as individually, is more and more through interlaced global markets. It is less and less through the blunt and divisive instrumentality of war, with the latter's attendant focus on raw territorial acquisition. The growth and success of the modern financial markets, as the premier example of how wealth and advantage are sought in the modern world, depends characteristically on the maintenance of a stable political base in those regions that house them. War and terror are inimical to such a base, as they render unpredictable the outcome of even routine economic decisions. Such doubt, in turn, and with a regularity that can be finely predicted, sends shock waves through the flighty equities markets, casting them into headlong dives. It makes such crises as 9-11 the rough equivalent of shouting into a peaked-out amplifier. It also makes potentially volatile regions, otherwise rich in investment potential, such as the former Soviet republics at the end of the Cold War, questionable places to invest, with (as indicated by their early and popular hedge fund presence) little to detract otherwise from their strong marketplace appeal. The harsh stigma of political instability artificially lowers such regions' economic value, leaving them at risk of becoming bystanders, rather than participants, in the ongoing economic bonanza.

Acknowledging the predictable impact of war and terror on market stability (with that stability itself now critical) gives developed nations, and in particular the most wealthy and influential elements within their borders, the incentive to not only safeguard economic development by not encouraging wars themselves, but by actively identifying and mitigating the sources of political and economic instability worldwide. Most typically, violence internationally can be attributed to, or is revealed to be fueled by, conditions of economic dissatisfaction or backwardness. This is an important general topic which will be addressed in more detail later on.

It had been true, of course, that America's entry into World War II helped propel the United States out of its Great Depression. It did so in a more decisive and effective way than either Franklin Roosevelt's charismatic leadership or his brain trust's pump-priming contrivances were independently able to do. While this would seem to indicate the existence of a general economic incentive for war, it is important to note how circumstances were different than they are now.

International economies were previously interactive, and strongly enough to spark a resonant downturn in both American and European fortunes during the immediate pre-war decade. This circumstance contributed significantly to Hitler's rise to power. Such markets were not, nonetheless, as seamlessly intertwined as they are today. Nationalism, as the impetus for war in the thirties and forties, and as expressed ideologically in its Nazi and Fascist extremes, has been modified, transformed, and mitigated in our own time by both the preponderant demands and overwhelming importance of international economic relationships. Such interdependent relationships have, in the current era, not only influenced national policy, they have essentially redefined the meaning and role of the nation itself. Coincident with this transformation has come the correspondingly enhanced role of the private transnational actor, the modern corporation, as a pivotal generator of wealth and employment, with as little regard as possible paid to traditional national boundaries. Beyond the turn of the century, traditional malignant nationalism has been largely defanged; it has been divested of the overtly belligerent shades it had exhibited in the 1930s and 40s. Open economic warfare by nation against nation has, at the same time, come to be regarded as being, in most instances, universally destructive. It is now perceived to be so by much the same logic as the use of nuclear weapons would have been, by those scorpions maneuvering gingerly in a bottle during the previous Cold War era.

Chapter Five

Into Unity and Light

When viewed within the context of humanity's long-term development, knowledge, technological or otherwise, is seen to play, not a negative or even a neutral, but a genuinely positive role; it is never knowledge, but only its incomplete and ignorant application which can be properly regarded as destructive. In keeping with the notion of an integrative world and of a larger, more integrative understanding, the expansion of knowledge in any one sphere should be viewed as part and parcel of an overall growth in understanding designed to lead us upward, and to unite the disparate members of our human community into a larger, more civilized, whole.

Nowhere is this process better exemplified than in the case of nuclear arms. Specifically designed as weapons of threatened or applied mass destruction, their development, as earlier outlined, brought about changes in the tenor of the world environment which defied the motives of their original creators. They forced adversaries to the bargaining table who otherwise would have met—perhaps even preferred to have met—on the battlefield. By barring the onset of full-scale war as a feasible political option, they institutionalized mechanisms of conflict resolution aimed at permanently avoiding it. Over the long run, the standardization of such mechanisms, and the continuous avoidance of full-scale war as a perpetually irrational choice, could hardly have had other than a positive effect on the development, East and West, of the competing societies. This remained true, despite the terror engendered by the overarching threat of nuclear war itself.

This effect has been particularly evident with regard to progressive developments in Russia. Continually subject to invasion during the course of its long history, that nation came almost impulsively to defend itself fortress-like from incursion by forces both military and cultural issuing from the outside world; this was when it was not, at the opposite extreme, embracing such cultural influences uncritically. With the attainment of nuclear parity, Russia, under Soviet rule, came to realize for the first time in its violent and tumultuous history a mode of security transcending the perils of its geography. New generations of Russians are becoming progressively accustomed to a previously unfamiliar climate of peace as their standard operating mode, and long-standing insular barriers are gradually, yet steadily, falling, though clearly amidst continual fits and starts.

Russia has too much of an authoritarian past for democratizing influences to transfuse its inveterate institutions all at once, and after the Westernizing reforms of Gorbachev and Yeltsin, ensued, under Putin, an inevitable, albeit restrained, Thermidorian reaction. Still, it is interesting to note, that Russia, no longer just "a feudal nation with an atom bomb" as it was once aptly described, is now at last emerging as a fully modern nation in its own right. It now exhibits much the same list of elevated priorities more characteristic of the advanced West. No longer fomenting violent revolution as a matter of course worldwide, it is now more likely to be concerned about its Chechnen terrorists (and the like) as a threat to its more conservatively-defined internal stability, and to articulate that concern through much the same invective as we have used to denounce bin Laden and Al Qaeda. This would remain true despite the fact that the frequent harshness of Russian policy toward Chechnya, both historically and more recently, would seem, at times, to place the shoe of random and inhumane violence on the other foot. Russia, despite any recent territorial diversions, has, as a trend, been further seeking to bolster, under conditions of relative stability, mutually-beneficial economic ties with the rest of the developed world.

In the final analysis, the advent of nuclear arms is comparable to other, more deliberately positive advances in technology, such as those in communications, which have had the effect, intended or inadvertent, of binding the world closer together. The dawn of the nuclear age thus marks a prominent threshold in the advancement of technology in its transformative impact on human history. This is a larger continuing process which is bound to precipitate still further changes and to foster an even greater interconnectedness of individual lives and collective destinies.

While a great threat currently is from nuclear proliferation and terror proceeding from less-advanced regions of world, regions which have yet to develop an indigenous nuclear capability, this trend too may be part of a greater evolutionary process implicitly aimed at enhancing our interconnectedness. The future development or current access to nuclear weapons by such nations, as a seemingly inevitable process, is likely to further compel advanced nations to address the problems of developing ones, so as to decrease terrorism and international crises from which all increasingly suffer, rather than viewing these developing nations as merely a source of cheap labor, or dumping ground for manufactured goods in a perpetually lopsided economic exchange.

As progress enhances interconnectedness, it becomes less feasible to deny to others what we would cherish and seek for ourselves. An isolated security, one which neglects the needs of others, is one which can be maintained only through the atavistic modality of violence, a modality unsuited to the changed conditions of the modern technological world. By effectively impeding other nations from advancing, we inevitably hinder ourselves from developing far enough beyond that outmoded mindset, thus jeopardizing our own and the planet's future. This brings us to the all-important principle of the moving inchworm.

According to the common analogy of the moving inchworm, with humanity essentially one, we can only hope to advance so far beyond the status of our least advantaged members before being constrained in our progress by their backwardness. Although the front end moves first, the back end must follow for the whole of humanity to advance in the direction we all intend to go. In the nuclear age, we are compelled to adapt to the ineluctable dynamic of accelerating change by advancing briskly into a new and increasingly integrative world. Our very survival depends on our so doing, and on our collectively refining, if not positively redefining, the behavioral norms that govern our routine interactions. Such a contemporary transition may be completed with relative ease by the most advanced nations, the United States among them. Yet the instability generated by others less advanced—the tail end of the inchworm—can drag all, even the most enlightened, back down to a level of conflict, belligerence, and violence at which such destructive norms are rendered more pernicious through the impact of ultra-modern technologies. The key assumption underlying the analogy of the moving inchworm is that the more-advanced and prosperous nations have a continued vested interest in making sure the less-advanced and less-prosperous keep up.

Bernard Lewis's observation from the 90s on the general American attitude toward the Muslim world is instructive in this regard. It, moreover, relates directly to what continues to be a key source of conflict in the modern world. As he stated in *The Crisis of Islam*:

> There is some justice in one charge that is frequently leveled against the United States, and more generally against the West: Middle Easterners increasingly complain that the West judges them by different and lower standards than it does Europeans and Americans, both in what is expected of them and in what they may expect, in terms of their economic well-being and their political freedom. They assert that Western spokesmen repeatedly overlook or even defend actions and support rulers that they would not tolerate in their own countries.

Relatively few in the Western world nowadays think of themselves as engaged in a confrontation with Islam. But there is nevertheless a widespread perception that there are significant differences between the advanced Western world and the rest, notably the peoples of Islam, and that the latter are in some ways different, with the usually tacit assumption that they are inferior. The most flagrant violations of civil rights, political freedom, even human decency are disregarded or glossed over, and crimes against humanity, which in a European or American country would invoke a storm of outrage, are seen as normal or even acceptable. Regimes that practice such violations are not only tolerated, but even elected to the Human Rights Commission of the United Nations, whose members include Saudi Arabia, Syria, Sudan, and Libya (105).

Despite what has been stated about the diminished role of conflict, the hard-learned lesson of the Munich Conference would seem to challenge the critical assumption that modern nations can attain peace by eschewing violence. As President Kennedy once observed in a 1962 televised speech, "aggressive conduct, if allowed to go unchecked and unchallenged, ultimately leads to war." The failed policy of appeasement has been cited as the immediate precipitating cause of the Second World War, and it indeed seems likely that had the British and French countered Hitler's preliminary threats, full-scale war would have been avoided. A number of Hitler's generals, terrified at his boldness, were prepared to seize control of the German government were his policies to meet with the anticipated reverses (Mosley 251).

Here the timidity of the former Entente may have contributed, not only to the start of World War II, but to its prolongation. The vindication of Hitler's belligerent policies, resulting from Britain's and France's failure to act, controverted the objections of his jittery generals and decisively stayed their hand. Even when the tide of war had clearly turned against Germany, the German people were still inclined, on the basis of this earlier vindication, to feel that someone in the German government other than Hitler was responsible for such troubling reverses. In the hope of stemming the tide of military losses, they put their faith in Hitler more than ever, and to the bitter end (Mosley 251).

Yet, in tracing the chain of events back even further, we can see that a lack of mercy, rather than an excess of it, may have been the deeper underlying cause of the war. Through the Treaty of Versailles, the Entente powers, at the end of World War I, exacted excessive reparations from Germany. Even more significantly, they saddled that nation with the now infamous "war guilt clause," blaming Germany for single-handedly instigating a war whose cause was at best unclear (Stearns 751). Amidst the rampant misery of a worldwide Depression and correspondingly high unemployment, the Nazis were able to exploit these collective irritants in their rise to power. While Hitler's aggression and the timid response to it were the immediate precipitating cause of the Second World War, the former Entente, through its actions prior to the war, greatly facilitated his rise. Different, more merciful actions may well have prevented it. In short, while the British and French at the outset of World War II were prepared to act in accordance with an agenda of peace, their own previous actions predisposed their erstwhile enemies to affirm that framework of violent confrontation from which, by then, there was no apparent escape.

The total lesson of World War II, combining the legacies of Munich and Versailles, is that there is, for the self-respecting nation bent on victory, no obvious means of thwarting the violent intent of a stalwart and determined enemy except by the use or threatened use of equal or greater force. Leaving aside for the moment the many cogent arguments to the contrary, the clearest way to prevent wars is to remove, as far as possible, the underlying causes which lead nations to turn to violence as a means of achieving their aims, and to keep the tail end of the developmental inchworm moving deliberately forward. We require for this purpose a broader definition of violence, one which takes into account the relative status of nations. The disenfranchised manifest violence by striking out against those who have more than they, the enfranchised by confusing attitudes of indifference and neglect with expressions of civility and peace.

In an increasingly interconnected world, the wanton neglect of our fellow man, amidst the busy feathering of our own nests, is itself a tacit consent to global violence. In a statement attributed to Edmund Burke, (though in an admittedly separate context), "the only thing necessary for the triumph of evil is for good men to do nothing" ("Edmund Burke"), and in the words of the Buddhist scriptures, "inaction in a deed of mercy becomes an action in a deadly sin" (Blavatsky 31). The contrast between the violence of rich and poor is largely one of their relative situations. The desire of the poor to promote change and to acquire status merely manifests more disruptively than that of the wealthy to maintain an unjust status quo.

Yet, to an equal extent, violent revolutionary movements, such as Marxism-Leninism, which have sought to instigate change through class or other modes of conflict, are repeatedly fated to engender only more conflict and violence as a result. As both Christian ethics and the karmic law of the East proclaim, means and ends are essentially one, and the methods used to promote social change become inextricably bound with and incorporated into the tenor of that change itself, and into the accruing social structure.

Lenin was perhaps the ultimate revolutionary pragmatist. A strong Nietzschean will to power melded inseparably with his revolutionary zeal. Nor was he hesitant to use violence to promote "peace," or to have the exclusive directives of a revolutionary vanguard serve as midwife to an all-inclusive workers paradise. Giving concrete articulation to the country's cry for "peace, land, and bread," he told the war-weary Russian masses in 1917 exactly what they wanted to hear. Through the Treaty of Brest-Litovsk, which concluded World War I in the East, he flexibly, albeit tentatively, ceded to the "capitalist enemy" in the form of the advancing Germans, territory severed from the Soviet homeland that with the invader's defeat was eventually returned to the Soviet orbit—directly (as with the Ukraine), eventually dominated (as with Poland), or "Finlandized," (as with Finland itself). During the period of the New Economic Policy in the 1920s, he led the Soviet Union to effectively retreat into capitalism as a means of promoting his socialist society's ultimate agenda, with the means, in each case, without any obvious loss, seeming to justify the ends. The problem going forward was that this unlimited flexibility of means to ends led his Russian state to ultimately, under Stalin, contort itself into something that by no person's definition could ever be described as utopian. Soviet Communism, itself, thus came to develop as one protracted stint in ideological self-deception, with the appearance of progress consistently counting for far more than its reality.

The greatest flaw in Lenin's, and hence, in the Soviet Union's seminal Marxist ideology, of its "dialectical materialism," was not so much its dialectic as its materialism. From a standpoint in results at least, Marx would have been better off leaving Hegel's more spiritual dialectic alone, for, as the *Bhagavad Gita* asserts from the standpoint of human consciousness, when we view things in a material way, we view them from the perspective of their potential divisions and conflicts. With ends and means truly part of one continuum, this leads in a far different direction than the integrative paradise envisioned.

While an integrative world *is* what we are heading toward, this goal is far better epitomized by such democratic principles as the United States has traditionally championed. It is typified as well by the non-violent political means of Gandhi, King, and Mandela, coupled with their notably more enduring results. It is typified by a toleration of differences, rather than their coarse eradication or suppression, and by an increase in routine interactions between cultures facilitated through enhancements in positively-rendered technologies. It is also typified by the continued fostering of the practical ties that bind, most notably in the economic arena. Notable is the fact that those nations that chose the alternate route to precedence and power of counter-evolutionary violence, however they may have prospered in the short term, were ultimately consumed by their own internal divisions and contradictions.

With the fall of France, Nazi Germany had effectively won World War II in Europe. Yet the domestic impetus behind foreign wars, as perhaps best portrayed by Orwell in his fictional *1984* (in which the ideal totalitarian state always has to be at war with *someone,* regardless of who it is, to justify its iron rule) served—at least in part—to prompt Germany, to inaugurate (after failing to exact any definitive agreement of surrender or neutrality from the beleaguered English in The Battle of Britain) a two-front war, by initiating an unprovoked attack against its erstwhile ally, the Soviet Union. This it blundered into doing, impelled, in no small part, by the internal imperative of absolute control. Another related component of totalitarian self-destructiveness involves the inherent structural weakness of dictatorship itself.

Hitler regarded the Russians as subhuman, and sought to obtain living space (lebensraum) and resources at their expense. While his attack upon Russia, at the time it was undertaken, blithely discounted the most basic tenets of German military strategy with its salutary dread of a two-front war, due to the despotic nature of the Nazi regime itself, none of Hitler's advisors was in any position to oppose it—or him. The brakes were off, and Hitler was allowed to lead Germany to ruin.

It is interesting to note that Germany's campaign to conquer Russia during Operation Barbarossa was greatly abetted by the matching flaws in the Soviet Union's own brand of totalitarianism. Prior to the German invasion, many of the most competent leaders of the Soviet military were either executed or consigned to prison camps. This dictatorial self-wounding severely jeopardized the Soviet Union in its capacity to respond to the German attack, and facilitated its enemy's early, dramatic victories. This eviscerating military purge was orchestrated under the dark impetus of Stalin's own far-reaching paranoia.

Such authoritarian systems, much as they may typify the ideal of short-term military efficiency, characteristically lack sufficient internal mechanisms for openness, creativity, and responsiveness to change required in the medium and long term to move the nation forward. The principle of the moving inchworm, and the way in which technology has extended the impact of developments from one part of the world to the rest, is further highlighted in the broader legacy of U.S.-Soviet relations.

The Soviet state that developed, first under Lenin, and progressively under Stalin, was as much a product of the harshness of Russian life, of the terms of existence in that part of the world, as of Marxist-Leninist ideology. As late as the 1930s, the United States could hope to remain isolated from developments occurring there, to the point of maintaining unchallenged, a false perception of what Soviet leaders were doing and of what the Soviet regime itself was like.

By the end of World War II and increasingly thereafter, the United States could no longer hope to remain isolated from the reality of larger world developments. At the same time, technology had reached a level which made it possible for the U.S. and the Soviet Union to go to war from opposite ends of the globe. While Hitler had designed bombers capable of extending his march of devastation to the American East Coast, their prospective development came too late in the war to have an impact (Hogg and King 87).

During the ensuing Cold War confrontation with the Soviet Union, America, for the first time, became susceptible to the worst kind of attack from nuclear bombers and ultimately missiles. Its position became comparable, though on a grander scale, to that of England on the eve of the Battle of Britain; increasingly as technology developed, neither channel nor ocean as natural defense barriers could serve to protect any nation from the threat of a full-scale attack. Nor could a policy of isolation, in an increasingly integrative world, any longer be realistically pursued.

The harsh nature of the Soviet regime was an indigenous development, a product of forces with which America was previously uninvolved, and, up until the modern era, relatively unaffected. Yet with the onset of the Cold War, advances in technology, and the ever-increasing interconnectedness of the world environment, America began to be shaped in ways not entirely positive by these previously isolated influences, in their continuing impact on the society we came to oppose.

In challenging Soviet totalitarianism under the aegis of defending freedom, we came, during the McCarthy era, to oppose it by methods more characteristic of the Soviets than of ourselves. In a truly interconnected world, where the perception nonetheless prevailed that violence could only be countered by violence, and threat by threat, we began to be reduced to the least common denominator within our collective environment, committing atrocities against our own citizens in the name of defending their rights.

Among the divided halves of postwar Germany, as a second illustration, communications technology maintained a cultural link that a wall set along a geographic boundary could not entirely sever. Those East German citizens who were able to surreptitiously monitor West German programs on satellite television—or more likely on shortwave radio—were repeatedly reminded of the disparity in relative living standards, making it difficult for the communist government to maintain the fiction that theirs was the path to progress.

Developments in technology likewise contributed to the demise of Soviet Communism. In the Soviet Union, as in its satellite nations, the myth of communist progress could be credibly advanced only through cultural isolation, by stifling the awareness of the Soviet population to events occurring in other parts of the world. Increasingly, with advancements in communications, and the corresponding emergence of an integrated world marketplace, the bubble of isolation burst and the deficiencies of the communist system became obvious and irrefutable.

A domestic example of the role of technology as a modern integrative force is the civil rights movement as it developed in the U.S. during the 1960s and beyond. That movement was influenced by technology, and by the progression toward an integrative world, in the following way: Technology created an international environment in which the interests of the United States came to be defined in increasingly global terms. Events occurring in other, distant regions came swiftly to our attention and ours to theirs, making the attitude of foreign nations toward the U.S. a continuous and immediate policy concern.

In such a world, the U.S. could hardly hope to function as a showcase of enlightened values abroad without practicing those same values here at home. This was particularly true of emerging nations whose populations were composed of groups which in America still existed as oppressed minorities. The advancement of the communications art, first through radio, and subsequently through the more dramatic medium of television, furthered cultural leveling, and put such states as Mississippi and Alabama effectively next door to those which had consistently disapproved of their civil rights performance. Fundamental changes in race relations to which the nation, as a whole, was hitherto resistant, even in the aftermath of a Civil War fought largely on their behalf occurred, as if by inevitable design, in the era of advanced technology.

Communications advances have also brought the values of the American middle class into the ghetto and intensified feelings of relative deprivation among its inhabitants. Advertisers, reaching an ever more diverse audience, have continually touted the virtues of a consumeristic "good life" which for many of the underprivileged remained, and still remains, out of reach. As demonstrated in 1991, by the videotaped beating of Rodney King, and, more recently, by the prompt and extensive Internet response to the Ferguson shooting, advances in technology have made the larger impact of a persistent inequality of status more pronounced. In making it more pronounced, it has also made it less tolerable overall.

Bernard Lewis makes a comparable statement with reference to our frequently strained relations with the Muslim world. As he observes, from *The Crisis of Islam*:

> The contrast with the West, and now also with the Far East, is even more disconcerting (*than the comparison to Israel*). In earlier times such discrepancies might have passed unnoticed by the vast mass of the population. Today, thanks to modern media and communications, even the poorest and most ignorant are painfully aware of the differences between themselves and others, alike at the personal, familial, local, and societal levels (117).

He further comments:

> The people of the Middle East are increasingly aware of the deep and widening gulf between the opportunities of the free world outside their borders and the appalling privation and repression within them. The resulting anger is naturally directed first against their rulers, and then against those whom they see as keeping those rulers in power for selfish reasons (119).

The more technology develops, the more interconnected our lives become. The more we enhance our interconnectedness, the less we can advance beyond the level of our least common denominator. By the tenets of realpolitik, as violence can only be countered by violence and threat by threat, the more we leave members of our common humanity behind in a more primitive state, the more we will feel compelled to quell their disputes through similarly violent means. The more events spin violently out of control, the more our responses, including the destructive misapplication of technology, will be reactively determined by them. Accession to modes of violence thus implies a loss of freedom.

The uncontrolled escalation of the Vietnam War, as a case in point, siphoned off brain power and resources which were to be allocated to the war on poverty. As a consequence of our Vietnam involvement, Lyndon Johnson, a man with vast legislative experience, one who dreamed of successfully implementing a comprehensive domestic agenda, is instead most readily recalled by posterity as a failed war president. It was the force of expediency, of a seemingly involuntary need to respond to conditions of conflict in a hitherto standard way, which likewise determined the earlier development of the atomic bomb. That development soon became a threat and source of concern in its own right.

In an interconnected world, the threat of nuclear violence becomes greater as a consequence of unequal development. It may be said, in this regard, that the United States, Britain, and Japan, among other modern nations, have entered the phase of integrative development. All have reached the limits of old-style territorial expansion associated with the nation-state in its primitive form, and are now focusing on the expansive development of commercial technologies.

Yet states like Saddam Hussein's Iraq (previously), and North Korea (currently), which had or have one foot on a lower rung of historical development, remain characteristically intent on chauvinistic expressions of military prowess. And they have avidly sought the technological know-how of the most advanced nations to accomplish their less-advanced ends. The gap between the technological means to which these polities have access and their overall level of development is broader in key respects than that of the original nuclear powers. As such, it represents a threat in some ways greater than that previously posed. While the potential for destruction is less than that associated with a superpower war, the probability of its occurrence is greater. Nations such as North Korea may not even have considered acquiring nuclear arms had more developed nations not acquired them first.

The crucial issue for mankind, well into the twenty-first century and beyond, is whether we will relate to the earth and to each other from the perspective of an integrated or divided world. The latter is what we have hitherto known throughout recorded history, and what is still largely dominant in the world today. The former is what we are progressively moving toward.

Through the efforts of some of the most inspired members of our collective humanity, we have penetrated into the quantum realm beyond the atom and its attendant Newtonian constraints. The Promethean fire of nuclear knowledge has, by a corresponding movement, been systematically lowered into the hands of far less worthy custodians. It has acquired full play at an atavistic level where it now represents a key threat to all life on earth. This remains true whether—with regard to the current climate of peace—that threat is immediate or not.

What we are now called upon to do to meet the demands of our collective development, is to complete our knowledge through wisdom, and to pursue the prescribed course of our natural and social development. We can do this by transitioning beyond a limited materialism and into a state of conscious spiritual progress and growth—a new dimension of relationships corresponding to the discovery of and transition to the quantum level. We may thus make a conscious part of our everyday knowledge and actions what had been previously consigned to the nether realms of mystery and faith. Our very survival demands that we do this rather than clinging precariously to a parsimonious Malthusian mindset and to the methods and motives of the past.

That we have, as a species, one foot in the light and the other in the darkness makes the current moment in our collective chronology particularly difficult and decisive. As with the "groundhog day" oil crises that continually come and go, yet with a distressing regularity inevitably return, the danger posed by nuclear arms may not be obvious and palpable to us at the moment; yet this only serves in the long run to render its hazards more considerable.

Chapter Six

The New World on the Horizon

The integrative international environment, the quantum-level world that we are now moving toward, is seen to manifest similarly in widely varied realms of human experience. Religiously, it is unfolding as a movement away from crusading conflict over competing conceptions of God, toward the development of a common religious base, one that integrates rather than divides, that sees value in alternative expressions of truth, and that appreciates like the facets of an exquisitely intricate and beautiful diamond the admirable diversity of all the world's great religious traditions.

This seismic shift in attitude is to be marked and measured as it proceeds by that most convivial of all human virtues—tolerance. With regard to truth, as Jesus himself professed, "you can tell the tree by the fruit it bears" (Matthew 3.16). Here truth is characteristically evaluated, not by means of its superficial expression in concepts or symbols, but through the substance of its living application.

True religion empowers people and engages them with life. False religion, as a purely negative force of societal control or oppression, disempowers and disengages them. Much the same may be said of the failed experiment in Soviet Communism, as a form of secular religion, in relation to its utopian origins and aims. False religion, as an intransigent force, is narrow-mindedly bigoted and riddled by fear—fear of life, fear of hell, fear of the material aspects of existence and of alternative expressions of truth. True religion is broad-minded, tolerant, and characterized by love. It intelligently discerns its own substance in the trappings of alternative forms, whether racial, societal, or cultural. False religion denies the value and validity of others' perspectives on the same living truth and, in delusively clinging to the conceptual shell alone, sees its own approach as final—its way as "The Way"—and all others as categorically false. In the new integrative environment now emerging, racial, as much as national and religious diversity, is not singled out and attacked, but celebrated and appreciated, ceasing to serve as the combustible and atavistic focal point for collective violence that it has more traditionally been.

As elegantly expressed by Hua-Ching Ni in his translation of the *I Ching*:

> In the last five thousand years of cultural development, numerous beliefs, customs and doctrines have been created and established that have either furthered or deterred the spiritual growth of mankind. The fact that differences exist should not be the "problem" that it often is in today's society. Instead, humanity should develop the understanding that these differences are only varying recognitions and expressions of different levels of spiritual achievement.
>
> Spiritually-developed people are greatly amused and entertained by these colorful differences. They can enjoy variation without becoming attached or confused, but such confusion has been the cause of many deep conflicts and wars among undeveloped people (591-2).

In international politics, the integrative state is unfolding as an inexorable movement away from belligerent nationalism, with its attendant focus on raw territorial acquisition and its establishment of a dominant chest-beating hegemon, toward a unified world in which national differences are properly celebrated and appreciated. Like the panoply of flags which adorn Olympic stadiums, the shades of individual nationalism ideally accentuate, within the broader integrative environment, diversity's intrinsic beauty, without, at the same time, highlighting the gratuitous overtones of nationalism's traditional violence.

Integration is further facilitated at the international level through the reinforcement of close and substantial economic ties. It is more pointedly expressed as a transition away from the Clausewitzian definition of war as an alternative instrument of policy, to the steady elevation of international competition to the more survivable realms of ideas, commercial enterprise, and material inventiveness.

With regard to Germany, as perhaps the clearest example of this ultra-modern trend, the very terms Prussia and Germany had at one time, and onward through Hitler, been virtually synonymous with militarism. The new Germany, still one of the most considerable of nations, is now, in keeping with its ultra-modern role, one of the most vital and supportive pillars in the new international economy. Its modern course reveals nothing to suggest that it will ever be motivated to revert to its militaristic past. Japan, as well, was once the most implacable of our enemies. The perception that the Japanese in World War II would have fought on to virtually the last man was a key determining factor behind the initial wartime use of the atomic bomb. Contemporary trends have bound the economies of the United States and Japan, making these formerly bitter enemies now the closest of friends.

With respect to the living environment, the integrative movement forward has manifested in a heightened perception of our oneness with nature, through which we seek, not only to survive, but to thrive by navigating skillfully the seamless web of relationships which constitutes the fabric of life, rather than by violently tearing through it. Whether with respect to global warming or to the peril of nuclear war, the need for an elevated consciousness and a sense of world unity have been accentuated now at a time when the consequences of our human vice and negligence have been exponentially magnified through technology's heightened role.

The heady pace of technological change has, as already noted, rendered us ill-prepared to cope with a nuclear-age environment in which finite actions can have virtually infinite consequences, in which decisions taken in the present can alter the future to the point of denying it. With nuclear arms, a mere test, particularly in the case of an air burst, has been known to release harmful radiation into the environment which may linger there for years. What impact past tests may yet have on the development of future generations is not precisely calculable. What is more confidently known is that radioactive residue, generated in one part of the world, can resurface in the food chain in another distant region. The seamless interconnectivity of our world environment and the loosely contrived nature of national boundaries are thus highlighted by a radioactive glow.

Strontium-90, a radioactive isotope contained in fallout, is known to be readily absorbed through the food chain, masking in turn as calcium, and resurfacing in concentrated form in cow's milk. Scientists, during the Cold War, routinely measured levels of Strontium-90 in milk to determine, after the fact, that secret tests or unpublicized nuclear accidents had occurred, so predictable and evident had been that deleterious signature. During the Cold War, populations in affected regions were repeatedly warned not to drink cow's milk for certain periods of time after nuclear tests (Shapiro). In the modern technological world, the essential unity of life, of human beings to each other, and of our form of life to all others, becomes ever more palpable, and our failure to acknowledge it as such, more evident and consequential.

While pursuing technological advancement, heedless of the human cost, has not yet led to the ultimate environmental catastrophe of thermonuclear war, it has already precipitated one nuclear and environmental crisis in Fukushima Daiichi, and another earlier in Chernobyl. This latter is a crisis which had been years, if not centuries, in the making, its roots threaded deep within the soil of Russia's history and consciousness. It is a crisis which has come to symbolize, both the generally imbalanced nature of modern industrial priorities, and the specific failings of the Soviet political regime. Serving as a warning for the future, it accentuates the need to acknowledge both individual human needs and the larger integrative environment.

Russia, long before Chernobyl, had been painfully aware of its technological backwardness vis-a-vis the developed West. This made the breakneck drive for modernization both a repetitive theme and elusive goal of successive Russian leaders. It was a priority highlighted in Lenin's description of Russian Communism as "Soviet power plus electrification of the whole country" (qtd. in Katzer). Yet, well before the twentieth century, its call was echoed in Peter the Great's drive to modernize his realm through the infusion of Western methods and technologies. Its impact was further evidenced in the colossal human cost of Stalinist forced collectivization, through which, as has been speculated, and as considered in conjunction with Stalin's purges, more Soviet citizens lost their lives than in the Great Patriotic War against Nazi Germany. While experts continue to debate whether and to what degree Stalin's extreme methods were the harsh medicine required to snap Russia out of its backwardness and to prepare it for the Nazi onslaught, the general consensus remains that his brutality was, in most instances, arbitrary and excessive, that it hurt far more than it helped, and that the requisite advances could have been as well—or better—achieved without it. The aforementioned purge of many competent military leaders whose absence was strongly felt at the onset of Operation Barbarossa supports this widely-held view. It is this same imbalanced drive to advance technologically, heedless of the human cost, with a Nature equated with an agrarian past regarded as something of an avowed enemy, that underlay the huge environmental mess left behind by the Soviets in their departure from Eastern Europe.

A balanced civilization, in the integrative state, does not proceed in this way. It respects both distinct individual needs and larger environmental concerns. It moves beyond the past by factoring the future into the present, thinking ahead to the ultimate human, social, environmental, and—then, of course, also—military consequences of its actions. National priorities change. Adversarial relationships change. In the transition from hot war to Cold War, German enemies and Russian friends became German friends and Russian enemies. The Mujahideen of yesterday, became the Al Queda of today. An Iraq armed against a dangerous Iran became a dangerous Iraqi threat, so armed. If we do not pay for our shortsighted policies with regard to such protean perils and alternating alliances, our children certainly will. We should not give them cause to curse us in disbelief, for behind the juggling act of shifting international allegiances, and tentatively-defined needs, stands the one, constant fact of our small and climactically unified world. It is a world where waters, clouds, and winds (whether the fresh winds of a sunny day or the tainted winds of Chernobyl) do not pause to reconsider their course as they stream across arbitrary and permeable national boundaries.

Nature thus continues to teach the needed, yet largely unheeded lesson of a united world.

No discussion of the movement toward an integrative world would be complete without considering, as well, the evolving role of women in modern society. In the United States during the 1960s, several integrative sub-movements proceeded simultaneously: the civil rights and environmental movements, as already noted, had begun, the movement for sexual choice, now reaching a crescendo beyond the year 2000 in the open (if reluctant) acceptance of gays into the ranks of the modern military, and the movement for gender equality. The latter may have unfolded without the violence which characterized the civil rights struggle, most particularly in the South, as blacks originally arrived on the continent as slaves and were visibly distinguished—stigmatized in fact—by their color, whereas women were subordinate only in relation to their husbands within the same family unit. Here the social devaluation of women may have been practiced throughout history, as a characteristic response, on the part of the just-as-characteristically prideful males, to the sure sexual dominance which women have possessed all along.

It has repeatedly been observed that sex is a game dominated by women. In the classical Greek comedy, *Lysistrata*, the women of the warring city states, by withholding their sexual favors, were successful in coercing their embattled husbands to put an end to a war of which they, as a gender, commonly disapproved. In the wily and surreptitious communiques that they (fictionally) exchanged across the battle lines, one can almost picture a modern counterpart in the verbose flurry of phone calls and impromptu lunches (the housebound female's equivalent of golf course capitalism) through which the cryptic rules of family life are discussed, evaluated, and set in stone—largely by women—before their stymied husbands get any solid inkling of what has actually occurred. But all this is prior to women assuming their proper and fully empowered role in society.

It is perhaps because of the underlying potential of women as a raw force of nature, and as the true "power behind the throne" in many families, with their influence as nothing less than an inverse expression of inherent male weakness, that men throughout history, in their overweening pride and longing for respect, have felt the need, it now seems, to relegate women to a subordinate societal role. It is almost as if, once the women were let out of the house, they would be destined to exert a superior power that could never again be restrained. Women's progress has thus proceeded, unlike the struggle for racial rights and the ship-ramming methods of Greenpeace, in a way as subtle, sharp, and effective as the women themselves tend to be—most relentlessly, yet without the blunt violence characteristic of cruder male-dominated methods and activities.

It makes perfect sense, as well, in this context that the women's liberation movement should have as its concomitant a revolution in sexual morals whose momentum has been sustained to this day. From the parade of increasingly risqué fashions (e.g., from Catalina bathing suits onward to thong bikinis) to the mesmerizing appeal of such musical divas (literally "goddesses") as Madonna, Britney Spears, and Katy Perry, whose performer status is associated as much with their sexual stage presence as with their musical gifts per se, it has uninterruptedly proceeded.

Rather than being repressed, at least in this one controlled arena of popular fashion and culture, the power that women possess, as rooted in their sexuality, is now overtly celebrated. It is likewise notable, in tracing women's emerging role in the integrative process that, as distinguished from the feminist movement in its nascent, tenuous, more self-conscious, and less secure form, as with past feminist objections to the beauty contest "meat market," women's sexuality and women's leadership are no longer viewed, at this current more developed and confident stage, as necessarily at odds.

With respect to the progress already realized on the road to gender equality, it is similarly notable, with the release of Jacqueline Kennedy's recorded diaries, that this intelligent, worldly and truly remarkable woman, one whose apparel choices dominated the fashion world of her day, who was notorious for stealing the limelight (as in her famous trip to Paris) from her almost ideally charismatic husband, a gifted woman with a keen sense of history, who translated erudite articles in French for her husband to read, and who was quite easily his intellectual equal, should have assumed such a deliberately obsequious role relative to him as something simply expected. Contrast this assiduously-cultivated self-deprecation to the conditions surrounding the contemporary role and rise of Hillary Clinton who, in being likewise the equal of her presidential husband, not only served as his trusted advisor during his period of presidential service, but emerged completely and independently from under his shadow as a savvy politico and efficient Secretary of State in her own right.

It is a matter of historical record that, while Western society was engulfed in the intellectual fog of The Dark Ages, and thus characterized at the time by Muslims as "barbarians," the Islamic world was lighting the way in such pivotal areas of knowledge as architecture, astronomy, and mathematics. A clear sign and convincing reason why so much of that formerly vanguard culture now lags behind on the road to integrative progress is the perpetual subordination, in many Islamic societies, of the fully one-half of their populations represented by women, women whose priceless talents are being routinely suppressed rather than productively utilized, as they have been more effectively (though as yet incompletely) in Western societies and Russia. This notable lack of female participation, in those regions where it persists, is both a contributing cause and telling symbol of non-integrative backwardness.

Within the realm of organized understanding, the momentum forward is expressed through an evolution in the scientific method away from exclusive reliance on a limited sense-based empiricism. It is transitioning from one in which ideas and experiences present themselves only as the inside and outside of Einstein's inscrutable watch[1] to an experiential domain in which the living basis of the truths behind our values is directly realized.

Like the meeting of the transcontinental railway from East to West, the convergence of modern science and faith is progressing apace to a stage beyond the current limitations of both. Such an evolution in our standard epistemology constitutes, at basis, a Newtonian to Quantum shift from the methods of the lower mind as they are identified in Hindu and Buddhist writings, to those of a higher integrative understanding. It is a gradual transition away from exclusive reliance on the standard approach to knowing and living by which the subject seeks to grasp its object in an effort to either understand or to acquire it, onward to the omniscience of the enlightened mind which, in being one with life, understands it most intimately, but which, as such, claims no ownership of it.

The latter, as a mystical attainment, identifies the concepts and creations of the material mind and the experiences of the sensory environment as one continuous flow in relative time. As such, it does not fall into the trap by which the contemporary world is tightly ensnared of reading absolute value into relative material relationships – relationships such as those which define separate national or religious identities. It recognizes such concepts as merely representing (re-presenting)—without containing—the more substantive truths and living realities behind them.

Such an integrative level of consciousness, rather than any material state to be realized through Darwinian conflict, is the true next phase in mankind's evolutionary development. Total knowledge and total virtue are thus ideally conjoined, and are seen to manifest as one in their individual and collective expressions. This is in direct contrast to the contemporary state and role of technology, in which the illumined understanding routinely serves the darkened ends of a harsh and counter-evolutionary divisiveness. Of this, the deployment and threatened use of nuclear arms remains the preeminent example. As described by Swami Satyananda Saraswati:

> Transcendental knowledge . . . is known in the form of intuition or illumination. The difference in this type of knowledge is that it comes from a totality of a situation. In other words, the whole arena of information is available and from this the answer is comprehended. In a sense it is like rational knowledge, but instead of a few facts, all the information is there to be utilized. This intuitive form of knowledge apprehends the totality of a situation: it sees the whole picture, nothing is missing. This comes from the superconscious realms of the mind during states of meditation. Rational knowledge is often warped by personal preferences and prejudices. Intuitive knowledge is independent of

all personal traits and projections. . . . Normally we are only aware of our limited, personal, and rational mind. This is the island rising above the sea. Yet beyond and deeper than this personal mind is the suprapersonal aspect of the mind, the seabed, from which all the islands arise. It is the realm of higher and more subtle vibrations which permeate the cosmos and existence. They are always present, yet during normal states of awareness they are not perceivable. During meditation a direct link is made between one's awareness and these higher domains of mind. From this comes higher knowledge encompassing everything that needs to be known. Meditation allows the deeper significance and nature of life and existence to reveal itself (226).

This evolution in consciousness, as it is quietly, yet inexorably, proceeding today is itself at the crux of the total forward movement, previously described, and as reflected in its merely subsidiary religious, societal, economic, and scientific transformations. Its sufficient attainment is our greatest—and perhaps only—hope, ultimately, for planetary peace and survival.

We have seen the opening spark of this transformation in the altered methodology of quantum science through which, as indicated by the Heisenberg Uncertainty Principle,[2] we now recognize the need to factor ourselves as observers into the context of what is observed. Still, acknowledging this quintessential relationship is a far cry from living it, or from fully incorporating it into our everyday responses.

In its currently more commonplace mode, a consciousness that seeks to grasp more complex concepts from a strictly separative standpoint, will continue to magnify precariously the destructive outcomes of the lower, "material" mind into whose service its discoveries and riches are now heedlessly placed. This is most vividly exemplified in the paradoxical dilemma of nuclear arms as a continued threat to our planetary survival.

In the commercial realm, the integrative consciousness is expressed in the transition away from a poverty-based Malthusian economics through which we define our material good in inverse relation to someone else's, as part of a narrow and competitive framework of finite gains and losses. This is meant to give way to a correspondingly quantum economic perspective in which the deeper truth and broader applicability of the Biblical story of the loaves and the fishes is fully and finally realized. By this elevated standard, the more we give and the more we serve, the more we ourselves can expect to receive and to benefit. The more we support the universal structure of which we are an integral part, the more we are lawfully supported by it, whether the evidence of this underlying exchange, the workings of the integrative or spiritual economy are immediately apparent or not.

The current economic structure is progressively evolving, nationally and internationally, from the oppressive domination of the weak by the strong, and the poor by the wealthy, a systemic imbalance that, for the bulk of the last century and a half, stoked the flames of Marxist revolution, and of a more generalized workers' unrest. It has progressed to the current, graduated stage of limited benefits secured through competitive outcomes and enhanced worker participation. This more recent phase is one which, nonetheless, continues to be characterized by a precarious imbalance between the haves and have-nots. It is meant to give way in the integrative state to the ultimate recognition of moral cause and effect as a law operative in the economic realm as fully as in all others.

We are now beginning to realize that, in the economic domain, as in our personal and national interrelationships, we have been living beneath our godlike potential, like Richard Bach's wretched seagulls, battling each other hatefully for the rancid fish carcasses and pre-gnawed bones scattered about the shores of our of material lives. We have mindlessly proceeded this way, all the while oblivious to the broader, more spiritual and fertile sources of individual and collective satisfaction available to us.

Such rules of the "spiritual economy" have been expressed in modern times as the law of manifestation. Its workings, as initiated from the higher plane of a more developed understanding, are dependent on a positive alignment with our total environment, and on the prior establishment, to some extent, of an integrative consciousness. This is an enlightened consciousness that is intrinsically aware of both its unique individuality and of its harmonious unity with the larger system of relationships in which its individual activities unfold. Its benefits are an expression of our essential oneness with life. This is a state attended by a true heartfelt feeling of unity with others, rather than by a need to dominate or suppress them, a quantum shift in identification which makes daily "miracles" of true abundance possible, while benefiting the individual in greater than material ways. This may seem to many like "fairy tale" economics, yet its workings would be no stranger than they are distant from, those of the quantum field being progressively mapped by today's developing science.

At the quantum economic level, we are beginning to discern that, in a substantially interconnected world, wealth is quite logically the karmic product of giving, and that the way of stinginess, if it does not lead immediately to poverty, will unquestionably produce a diminished total environment of enjoyment in which all collectively suffer. Here the wealthy nation or individual, without being simultaneously virtuous, will be seen to be unable to deeply, permanently, or comfortably enjoy what was so tenuously secured through so much strife and hardship.

As the Buddhists proclaim, our individual capacity to appreciate even material things is directly dependent on our intrinsic level of virtue. It has been said, in this connection, and within the context of the Buddhist cosmology, that when a human being tastes water it is water, when a god tastes it, it is nectar, and when a hungry ghost tastes it, it is fire and pus. As with the Heisenberg Uncertainty Principle, what we perceive is dependent on who and where we are as participants. That status is, in turn, dependent on the operation of cause and effect, on the outcome of our individual and collective choices.

This observation furnishes an interesting and potentially revealing answer to the age old conundrum once posed by John Stuart Mill as to whether it is better to be Socrates dissatisfied or a pig (or fool) satisfied. The former is clearly indicated, as Socrates could infer a wondrous value from even the context of his death's troubling circumstances, whereas the quality of the pig's enjoyment is darkened and constrained by the limitations of its far narrower nature. It is the task of humanity—both individually and collectively—to expand our nature beyond its current limits, and to thereby partake of the nectar rather than of the dregs of life, attaining, by such means, the true satisfaction and freedom from suffering that we each inwardly seek, and that we all now collectively require.

Last but not least, in the realm of modern man's tragically divided psyche, we are coming to recognize, not only the holistic interrelationship of physical to mental, as already routinely acknowledged by modern medicine and psychology, but the critical importance of reintegrating the spiritual dimension of human experience back into a totality with the rest—with "psyche" itself finally restored to its proper original meaning as "soul," and with the soul reclaiming its central governing role in our lives.

This transformation can be expected to occur through evolutionary progress, and from the ineluctable recognition that mankind cannot be happy, let alone fully sane or developed, without this most distinguishing element of human nature—the spiritual—properly re-factored in. The resulting understanding is not one tragically fated to be relegated to the realm of mystery or authority, and thus placed tantalizingly outside the purview of the individual's immediate experience, just outside the way we typically apprehend what we currently claim to know. We simply need to become accustomed to new ways of thinking, being, and knowing for such a perspective to be recovered. And more than being positive in our beliefs, in becoming spiritually centered, we will learn not to be bound by our own previous constructs as we come to identify progressively with the essence of who we are, rather than with the collection of conceptual labels we have scattered across our minds, and through which we artificially divide ourselves from others and from the rest of our living environment. We can thus entertain the hope of nimbly traversing the landscape of varying points of view on the road to an ever-higher and more-integrative understanding. We will then come to use our calculating minds as the essential tools they are, as adjuncts to the spirit's higher governance, rather than misapplying them to self-destructively extend the imbalance of an unwise knowing into avenues of further jeopardy.

This rediscovery of the spiritual dimension of human action and of elevated human experience is the real underlying meaning behind Plato's cave allegory, which essentially refers, not merely to obtaining a predefined level of conceptual knowledge or intellectual satisfaction, but to breaking through to a radically new and superior way of knowing in which that which understands and that which is understood are apprehended together (Buchanan and Jawett). The resulting condition is one which, throughout recorded history, has been comparably identified in various religious cultures—that of the Sufis, the Buddhists, the *Bhagavad Gita* and the Christian Mystics, among others—as a state of enlightenment. It is a state which the "modern" world, preoccupied as it is with its own superficial displays of sensory glitter, with its own material shadows, has temporarily lost sight of it. It is nonetheless resurfacing as a rising tide which is newly reasserting its pull, lapping around the edges of today's popular literature. It is one which, provided we survive the consequences of our stubborn collective ignorance, we seem destined, in time, to reclaim.

This discussion of the soul leads logically to a consideration of the important question as to where—if anywhere—our human advantage resides. In what does it essentially consist? In proclaiming "I think, therefore I am," philosopher René Descartes seemed to suggest that our humanity is essentially defined by our ability to think discursively. Yet current events have already shattered any value to be attached to this traditional assumption alone.

On March 11, 1997, a key threshold was reached, though perhaps unnoticed by many at the time, as world chess champion Gary Kasparov was defeated by his cybernetic rival Deep Blue. Here a machine was able to surpass the most capable human opponent in a key domain traditionally associated with intelligence. The cyber-evolution underlying Deep Blue continues relentlessly at a pace far exceeding the current limits of our human, physical adaptability. We cannot, at present, transform ourselves with anything resembling the flexibility by which we routinely develop our technologies. Our wounds heal only with time, while machine parts can be replaced instantaneously.

At the same time, any effort to bridge this expanding gap through eugenics, is a path known to be fraught with moral peril—if not with absolute evil. By identifying and attaching importance to what is "better" in human traits, we cannot help but, by indirection, single out what is "worse"—what is less or least desirable among human beings themselves.

It may very likely be that our human limitations are, as considered from a more enlightened perspective, the very fuel for growth, in the highest octane imaginable, needed to propel us spiritually forward. Like the gravitational force that must be overcome to catapult a spacecraft beyond the earth's orbit, our material impediments serve as the resistance required to accelerate our souls' development in a way that cannot itself be materially defined or evaluated.

We gain spiritual strength through overcoming obstacles and accepting challenges that enable us to unfold our inner divinity. To aim at evolving on simply a material level is thus to miss the very point and purpose of human existence. Any obvious limitation, any handicap physical or mental, can be seen in this context as but a barbell in the soul's time-space gym, conducing to strength of character in the dazzling will to overcome.

As viewed within this evaluative framework, the runner's prize is at basis a triumph of stamina and spirit, not a means of measuring his speed against the cheetah's. The chess player's is one of psyche and will against those doubts and inner weaknesses that would plague him and threaten to rob him of his personal victory, not a merely robotic exercise in mechanical thought. Within the context of such human attainments, as they are ultimately and spiritually defined, the Olympian and the Special Olympian have equally as much of which to be proud. As Mike Dooley notes in *Choose Them Wisely: Thoughts Become Things*:

> Life is a "package deal." If today you were a bit wealthier, skinnier, smarter, or cuter than you already are, there might have been prices to pay in other areas of your life. Maybe you wouldn't be as insightful, spiritual, or compassionate as you are today. Or maybe you wouldn't laugh as much, help as much, or roll exactly the way you do now.
>
> Get the picture? The events and circumstances, challenges and accomplishments that have shaped your life, that have left you wanting more of this and less of that have also honed your finer senses and polished your character as nothing else could have (118).

Nowhere do we find ourselves more at the crossroads than in this particular domain, for here we are presented with a critical choice between a humane and inhumane future in how we define ourselves and our progress. We can, out of continued fear and competitiveness, obsessively develop our soulless material technologies while neglecting to develop ourselves. This would translate into a brakeless deterministic progression that establishes our defining role merely as midwives in the ultimate development of super-rational machines, machines that must, if we adhere to such a course unswervingly, inevitably supplant us. This fateful drive is likely to be supplemented by an equally obsessive effort to artificially bridge the ever-widening gap separating human and mechanical capabilities, through such eugenics programs as would make Hitler's plans for a master race appear comparatively paltry, or by bizarre and monstrous attempts to mate human and machine parts.

The road leading to this dystopian end would invariably be marked by a pervasive fear and competitiveness that would, like a virus usurping the workings of a cell, lead us to develop our commercial and military technologies at an ever-accelerating pace, heedless of the human consequences. Our fellow beings would serve mainly as fodder for this purpose, much as the infantry of the past once served as "cannon fodder" in traditional wars. If this scenario sounds at all familiar, it should, for it is basically where we are today. We routinely use and dispose of each other for the attainment of goals—personal, economic, social, or political—worth less, by far, than our human connections. The alternative, more beneficent approach is to open the unseen "roof hatch" to our soul's natural and compassionate evolution, short-circuiting through a progression of values and morals, the current misdirection of our technological advance, one that, in its current wayward form, can only lead ultimately to our collective demise.

Our essential human greatness lies in our souls. This is a greatness we are meant, each in his or her own time and way, to discover, and which the ordinary constraints of our material lives ironically do the most to unveil. Still, if it is not to be our ultimate fate to just dwell dumbly among our scientific creations, whose growing sophistication is driven by those competitive forces that threaten to ultimately consume us, we should explore, along with the soul's heightened purposes, its potentially unlimited scope and powers. Even now, we occupy to an ever-increasing degree, a world of objects, the workings and sophistication of which lie beyond the pale of understanding for 99% of our ken. In vacant lots, steam shovels and cranes sit dormant like slumbering dinosaurs, having already surpassed us in raw physical strength, but currently only as brainless wonders, silently awaiting the implantation of the cyber-intelligence required to complete them.

Should civilization be sheered down to nothing, through a nuclear war, culturally-eviscerating plague, or similar manmade or natural catastrophe, how many of us would be able to rebuild even as mundane a device as a radio from the knowledge he independently possesses? So much of what we personally enjoy is the culmination of many minds' efforts, much as the computer that defeated Kasparov, combined, in effect, and successfully engineered into a single intelligent platform, many of the best players' strategies, styles, and most-favored moves, including those of Kasparov himself.

Beyond the backdrop of any material limitations and troubling reflections on our technological future, lies the largely untapped resource of our soul's dormant powers, currently cloaked in pedestrian garb and buried beneath the merely mundane and mechanical pursuits of which we are now unduly obsessed, and which occupy most of our time. In our spiritual natures, rather in our physical abilities or analytical skill alone, resides the potential, largely unactualized at present, to remain head-and-shoulders above our physical creations, and in continued alignment with standards that remain compassionate and quintessentially human.

A calculating machine can never go wrong, for it can never imagine things other than the way they are. Its tumblers fall into place mindlessly, and hence unerringly. It is our human capacity to imagine that both permits us to err and empowers us to create a better future. In encompassing both of these extremes at once, it illuminates, perhaps better than any other faculty, and far better than reason alone, what it truly means to be human. By envisioning more, we attain better, as we advance toward a progressively integrative world.

As we proceed toward that integrative world, we shall hopefully realize in time that to see life from a material vantage point only is to view wholeness from the perspective of partiality, a partiality whose unity is but the drab uniformity of grey goose-stepping troops, and whose built-in concomitants are fear, division, and conflict. Yet to approach physical existence from a spiritual angle is to view what's partial from the perspective of what's whole. It is a viewpoint whose collective realization involves nothing less than a radical transformation of current conditions to an integrative world environment, one in which the soul's native dimension is rediscovered and reclaimed, and its dominance over material conditions, pragmatically realized. It is a return, in effect, to the legendary Garden of Eden.

NOTES

1. "Physical concepts are free creations of the human mind, and are not, however it may seem, uniquely determined by the external world. In our endeavor to understand reality we are somewhat like a man trying to understand the mechanism of a closed watch. He sees the face and the moving hands, even hears its ticking, but he has no way of opening the case. If he is ingenious he may form some picture of a mechanism which could be responsible for all the things he observes, but he may never be quite sure his picture is the only one which could explain his observations. He will never be able to compare his picture with the real mechanism and he cannot even imagine the possibility of the meaning of such a comparison" (Einstein and Infeld).

2. The Heisenberg Uncertainty Principle can be summarized as: "The notion of the observer becoming a part of the observed system is fundamentally new in physics. In quantum physics, the observer is no longer external and neutral, but through the act of measurement he becomes himself a part of observed reality. This marks the end of the neutrality of the experimenter. It also has huge implications on the epistemology of science" (Knierim).

Chapter Seven

The Law of the Jungle
and the Return to Eden

The primitive struggle for existence, as authoritatively described by Darwin, was not, in essence, a struggle of life against life for absolute dominion of the living environment, as have been wars among nations in modern times. It was, at basis, a struggle within and among diverse species for the control of scarce resources needed for basic survival. As Darwin, himself, described it:

> This is the doctrine of Malthus, applied to the whole animal and vegetable kingdoms. As many more individuals of each species are born than can possibly survive; and as, consequently, there is a frequently recurring struggle for existence, it follows that any being, if it vary however slightly in any manner profitable to itself, under the complex and sometimes varying conditions of life, will have a better chance of surviving, and thus be naturally selected. From the strong principle of inheritance, any selected variety will tend to propagate its new and modified form (44).

In the modern world of vast and accessible resources, primitive style conflict, using the sophisticated tools of the modern technological environment, serves not so much to enhance the survival of the human species, as to endanger it. It does not ensure the acquisition of scarce supplies, but rather the loss of abundant ones. The irony is that civilized man risks through war the loss of what he already has in peace—an ample supply of all he really needs, or the ready means of acquiring it.

Full-scale wars, in creating artificial conditions of scarcity in the very midst of abundance, lock mankind into the more rudimentary struggle to survive. They create, rather than true security and abundance, an anarchic world of rampant mutual suspicion in which life, as expressed in Hobbesian terms, is "nasty, brutish, and short." By choosing war over peace, violence

55

over civilized norms, modern man chooses to regress. On this basis, it can be hypothesized, with the adaptive response rising to meet the evolutionary challenge that humanity is at the crux of a progression from one historical and evolutionary phase to another, with the wake up call to the higher phase occurring now.

The lower phase, half-buried in the animal realm and characterized by continual violence, corresponds to the struggle for scarce resources described by Darwin. Its reflex is to be found in the repeated incidence of war; the higher phase, of which we are now becoming progressively aware, is characterized by the full actualization of the human potential under conditions of abundance. It is a phase at which humility is found to be more conducive to growth through understanding than a belligerent pride, and compassion more instrumental to collective survival than violence.

It is worth noting, in this regard, that the principles of the great religions, principles which reflect the highest in human nature, are for the most part the opposite of what it takes to survive in the animal world: The law of kill or be killed versus love thy neighbor; dominance through force vs. turning the other cheek; pride vs. meekness; looking out for number one vs. the principle of the good Samaritan; the longing for peace vs. the perpetuation of conflict as a reputed catalyst of evolution; grasping vs. giving.

Yet there is every reason to believe that these higher principles make for harmonious individual lives and social orders. Their opposites, particularly in a world of high-tech weaponry, make for lives of despair, meaninglessness, fear, and ruin. They appear, in the contemporary era, not only profound but prophetic, for they have more direct meaning now, in the nuclear age, than when they were originally uttered. It is as if mankind, as a species, needed all these many years to fully learn them. The status of our collective future is thus to be determined, and determined here and now, by whether we stake our claim through the higher, specifically spiritual, laws of abundance, or atavistically and precariously descend to even darker depths within the more commonplace norms of conflict and scarcity instead.

It would be worthwhile, at this point, to pause and to consider why these two states should be, in fact, so diametrically opposed, yet each so obviously true within its own separate sphere of life and activity. Why would—how could—such sentiments as love and universal brotherhood—sentiments so distinct from what we have hitherto known within the primordial scheme of nature, and in the historical legacy of mankind, be just as natural and just as true—and even more so for us moderns—as contention and brutality? So opposed in fact are these conflicting modalities as to make the Tertullian notion of believing because it is absurd appear almost sound, with that absurdity being the only conceivable basis for optimism. Yet such ostensibly opposed states are, in the final analysis, both as natural and as true as the shifting of the tides, the changing of the seasons, or the transition from night to day, all depending on who, what, and where we are.

In the first evolutionary phase, within the realm of primitive nature, life prevails by adapting successfully to the conditions of the material environment; in the second, by demonstrating the supremacy of quantum or spiritual truths over material constraints and challenges. In the primitive and strictly material phase, any obstacle we might encounter would be perceived as just that. The obstacle itself would be narrowly despised, the person or circumstance embodying it, unceremoniously, and perhaps even violently, squelched. In the second phase, that of our progressive spiritual liberation from the constraints of material bondage, physical obstacles, whether manifesting in the form of outward circumstances, inner shortcomings, or harsh and inimical people, are transformed from sources of contention and loathing into efficient means—and perhaps the most efficient means of all—of developing spiritual strength. They may be, and often are, indistinguishable in form from those more ordinary material challenges with which we are all-too-familiar; but they now assume, in this latter stage, an altered and elevated significance. We may not welcome them, and, in not being masochists, generally will not seek them; but we are far more likely, when spiritually mature, to acknowledge their worth, extract their lessons, and utilize them as the efficient means they have alchemically become of developing the compassion, patience, and ethical skill which are the hallmarks of a higher existence. These are the same factors which we increasingly need on a global scale to survive.

The fundamental difference between the "trials of life" and the "trials of the saints" consists almost exclusively in the environment, spiritual or material, to which the organism is compelled to adapt. Material life prevails by rising superior to the obstacles imposed by material conditions, spiritual life, by overcoming obstacles to spiritual realization imposed by the material realm itself. In the material environment, physical survival is the sole and pertinent end in view. In the quantum environment, the altered and elevated aim (as emphasized by Gary Zukav and Linda Francis) is spiritual growth (Zukav and Francis 24). As Patanjali similarly observed, in relation to this higher striving, the purpose of Prakriti (the material world of sensible objects and experiences) is the liberation of the Purusha (the soul or spiritual element within creation) (as qtd. in Carrera 242; 171). When we factor in this single, yet critical difference, the difference in focal environment, all of the earlier seeming contradictions readily resolve themselves. All of the separate and scattered pieces fall meaningfully into place.

In the struggles of nature, the victory achieved is the relative victory of material survival over death, i.e., the realm immediately below, over the condition and lesser status of inert matter. In the spiritual realm, the victory is one of spirit over the constraints of a narrowly materialistic life—the realm immediately below that; in this latter phase, each victory won is a victory over, rather than within, the constraints imposed by that earlier, material phase; each victory is a triumph of the soul. Failure in the struggle for material survival results in material death, in the ratcheting down to the level of inert matter. Likewise, as the Bible proclaims, "the wages of sin is death" (King James Bible, Romans 6.23), not death in any obvious physical sense, but death as a spiritual being, one afforded the precious opportunity to live by a higher, more advantageous, more subtle—yet equally real—set of favorable standards and laws.

When one dies spiritually, he lowers himself, though sin, "vibrationally," thus placing himself, as did the expelled couple in the Garden of Eden, under the dominion of those harsher laws and energies associated with material struggle, compelled, in matchingly contentious environments of scarcity, to focus narrowly on the physical, living in perpetual fear, for survival ends alone, and "by the sweat of one's brow" (New International, Genesis 3.19). For a spiritual being, one who cannot truly find any meaning in the cutthroat ways of savage Nature alone, that is death indeed. What Darwin observed in the animal realm was the cresting of the divisive tide, in the movement of spirit into matter. The ebbing of that tide is the return of spirit to its Source, in the Living Universe, or God. Both are separate, yet ultimately consistent, components of a grand evolutionary scheme.

As Swami Saraswati notes, the fine tuning of relative advantages in relation to the outward environment, the whole phase of material evolution that Darwin so masterfully described, continues (196-197). But having exclusively to do only with our becoming rather than our being, it can have no satisfactory end, no absolute value, no enduring meaning in itself as viewed from the finer vantage point of the higher evolutionary ground. It is but a Ferris wheel that can only go round and round without finality, producing at best a transient rainbow of pyrrhic victories whose lamentable finality is inevitable. Once this is clear, it becomes excruciatingly clear, challenging our limitations, and demanding a corresponding transformation from us.

The error we are most likely to commit as we first attempt, without sufficient judgment or experience, to calibrate our lives to the demands of the higher phase is, not in continuing to acknowledge the worth of relative material advantages, or even to demonstrate a natural and practical preference for them. It is rather to regard the relative as if it were itself absolute, to confound the former with the latter, and to give the former's shifting standards precedence over the deeper Truths of the higher domain. To regard the relative, the limited, and the material as if they were, in effect, God, was the Biblical sin of the golden calf, a collective error in judgment attesting to a falseness and narrowness of perception, one that would be attended by far greater consequences now, than it had then.

It is never wrong, for example, to prefer wealth to poverty, but it would be to see money as being intrinsically worth more than people. It is not wrong that we prefer to go to the prom with the stunning young woman with the blond hair and blue eyes, or to regard her as being particularly beautiful; but it would be to think that the world, as a whole, and as based on our personal preferences, should only be inhabited by blond-haired and blue-eyed beauties. It is never wrong to note the evident value and worth of physical strength and intelligence, and to naturally admire those qualities over dense stupidity and weakness; but, it would be to assume that those who lack such favorable traits should be viewed as less valuable than those abundantly gifted with them. It would not be wrong for us to love and to appreciate any one separate part of creation, any possession, any form, or any cherished individual; but it would be view that part as if it were the whole itself, and, in venerating it as such, feel utterly incapable of existing without it. It is not wrong that, as grateful citizens, we should love and serve our country, so long as we do not, in the process, hate others who also love their country, thereby violating the larger principle of planetary unity. As inhabitants of a spiritually-trending age, we are obligated by the natural precedence of spiritual over material to wisely choose the former over the latter, where the two may happen to conflict.

It is telling that, when we look closely at what constitutes sin and evil in the great religious traditions of the world, it is found at basis to be none other than for the spiritual being to behave in a strictly animalistic way, with the Devil himself frequently portrayed as a degraded being with hoofs and horns. This perspective should not be regarded as a mere Freudian denial of our true, yet contentious, nature buried beneath society's unduly restrictive norms, but something more broadly natural, more in keeping with the way we are genuinely evolving. Anger and aggression, fear and greed, are all atavistic expressions of primal territoriality, whether demonstrated by individual men or nations. Their unfiltered expression is, more and more as we advance, attended by pain and consequence, rather than by any reliable happiness, and this for the very reason that it is limited and wrong for us. That we can already sense this deeply, and witness its evident impact on our lives, is what makes "faith the substance of things hoped for, the evidence of things unseen" (King James Bible, Hebrews 11.1), or, in Pascal's terms, gives the heart "reasons, which reason does not know," (277) and this despite all of the world's apparent harshness and turmoil.

The crucial dilemma facing the modern world is that, with our vastly enhanced knowledge and capabilities, we have gone only part of the evolutionary distance we are meant to travel. We too often use the higher only to magnify the lower, the magisterial products, power, and tools of the inspired (i.e. spiritualized) intellect only to digressively serve the selfish ego's ends.

With the enhancement of that disparity comes the increasing potential, in the modern technological world, for unmitigated ruin.

Still, today's average citizen might well be inclined to exclaim, "Yes I believe in God, in good vs. evil, and in the importance of spiritual values, but, you know, this is the 'real world,' and you must claw your way through and compete, undermining the interests of others in order to survive." Even organized religion has been guilty of tacitly buying into this underlying premise by construing heaven as something attainable only after death, while the world we live in is effectively written off as an ethical wasteland, hopelessly overrun and dominated by the ways and tools of violence, deceit, oppression, and evil. In so espousing such a view, however obvious the outward conditions may be that appear (and only appear) to confirm it, we proclaim our lack of familiarity with the scheme of life as it is genuinely being enacted in our more advanced and pivotal age. That in this higher and greater time, Love is Truth, and peace, a universal need, is indeed the "Good News," not only of Christianity, but of every higher religion that in its own unique and valuable way expresses this same sentiment, with the wages of sin, in the post-nuclear age, to a far greater degree, still death.

The lofty and explicit goal of Milton's *Paradise Lost* was "to justify the ways of God to men" (line 26). This book's aim is, more modestly, the same, in its attempt to reveal the higher evolutionary course through which the world is genuinely—and pragmatically—evolving. It is intended to show that evil isn't designed by nature, or Nature, to triumph, and that, in actual fact, the ways of virtue are woven into the very fabric of Life. Even in this uncertain world, where conflict and war are seemingly so ubiquitous, virtue remains, in the final analysis, the ultimate pragmatism, whereas the force of evil is like a wantonly destructive wind that, in the long run, must cannibalistically expend itself. It is known to do so, however, only after tragically obliterating from the landscape of human experience, and from the experiential range of the future's breathless enchantment, so much in life that is beautiful, true, and good, in the loss of those priceless specimens of art, architecture, and culture—not to mention all the lives cut short—that will never be observed by loving or wondering eyes again. This is a cost, even if it is eventually surmountable, that should never be taken lightly, one that this book strenuously encourages us to forgo.

In the modern world, "Love thy neighbor as thyself" (Mark 12.31), has become, not just the individual door to heaven, as it is has been most generally perceived in the past, but the collective entry-point to the most rapid progress of civilization itself, and, in the post-nuclear age, to humanity's future survival. The sufficient absence of this principle can only spell our certain and inevitable doom.

Mankind's covenant with Nature, as embodied in such statements as the Sermon on the Mount, would seem to be fundamentally different from that of the brutes; one is expanding within the set limits of its physical nature, the other progressing beyond them. The more we develop beyond the constraints of the natural world, the less instincts rooted in the conditions of that world have their part. That our greatest obstacle, in the course of our evolutionary development, is, not the environment which we have helped to shape through our inventiveness and creative will, but our vestigial instincts and primal nature, suggests how the timely call to awakening may be trolling now, as annunciated by, and articulated through, the survival demands of Nature Herself.

Man, by centering his consciousness on the communal values of his higher nature, in developing the character needed for civilized life, ultimately prospers in creature comforts as well as in ideals. In making his violent instinctive nature the focus, he denies, not only his spiritual well-being, but ultimately, in a world of nuclear and other weapons of mass destruction, his physical survival as well, plunging back into a violent and dissatisfying realm of anarchy and scarcity instead, a realm to which such rougher instincts would more aptly apply.

Yet if this framework of spiritual evolution is, in fact, true, then why is it so hard to see? Why do such theories as traditional Darwinism continue to prosper to this day, extending the tentacles of their misapplied influence into the ambient realms of our commercial and social, national and international interactions? Could not our inability to see this theory's limitations, as a strictly materialistic doctrine, in itself be the result of a previous epic fall into blindness? Might Genesis not be correct after all, though hardly from the standpoint of any traditional banal argument of Creationism versus Evolution, but in an alternative way not out of keeping with the dictates of a scientific understanding? And after further consideration, may we not be left to ask whether Darwin's theory, in the specific way that it pertained to the brutes, ever have fully pertained to civilized Man at all?

Correct in its own sphere, as Newtonian physics was and remains in relation to Quantum Mechanics, it leaves an important part of the picture for mankind inadequately—and thus misleadingly—explained. If this is indeed the case, with a need in evidence for a modern corrective, then why do elaborations and extensions of that theory, as currently misapplied, maintain such a hold, even on people who claim to be religiously inclined?

Here Buddhism, being more experiential than contemporary Christianity, while embracing many equivalent values, may well provide the "missing link" (pun intended) between science and faith. It would serve to reconcile a limited empirical view which focuses on physical evolution alone, seeing conflict under conditions of scarcity as the catalyst of evolution, and a Spirituality that touts love as the supreme and most highly-developed of our intrinsically human virtues. Here the transcendent value of love would seem to pertain in a universe that, in the final analysis, is both benevolently intelligent and from the unique human standpoint, appropriately abundant.

The entire realm of material change to which the natural struggle relates, and which the traditional form of Darwin's evolutionary theory encompasses, corresponds to what the Buddhists term "samsara." Samsara, itself, maps directly to the material realm of ongoing change and its activities. Should such activities not have, as their reference point, something higher, they are inevitably revealed to be essentially empty, transitory, and meaningless, and particularly so for human beings, who have a definite spiritual core to their natures. Here the conditions of what has come to be termed "the natural struggle" resonate vividly within the Buddha's youthful reflections on life, reflections which once they had ripened into a dominating weltschmertz propelled him to seek enlightenment. Quoting from *The Light of Asia*:

> About the painted temple peacocks flew, / The blue doves cooed from every well, far off / All things spoke peace and plenty, and the Prince / Saw and rejoiced. But looking deep, he saw/ The thorns which grow upon this rose of life: / How the swart peasant sweated for his wage, / Toiling for leave to live; and how he urged/ The great-eyed oxen through the flaming hours, / Goading their velvet flanks: then marked he, too, / How lizard fed on ant, and snake on him, / And kite on both: and how the fish-hawk robbed/ The fish-tiger of that which it had seized;/ The shrike chasing the bulbul, which did hunt / The jewelled butterflies; till everywhere / Each slew a slayer and in turn was slain, / Life living upon death. So the fair show / Veiled one vast, savage, grim conspiracy / Of mutual murder, from the worm to man, / Who himself kills his fellow. . . ,/ The rage to live which makes all living strife—The prince Siddhartha sighed. "Is this," he said, / "That happy earth they brought me forth to see?" (Arnold).

The outcome of the Darwinian conflict, "the rage to live which makes all living strife," is forever tentative. Its "winner" thumps his chest in blind arrogance atop a foul dung heap of impermanence. Fearfully grasping his temporary prize, he inherits with it the all-but-certain knowledge that his situation too must ultimately deteriorate, and any limited material advantage relinquished to the latest benefactor in the repetitive struggle.

It is worth repeating, in this context, that Darwin's theory is not a value system. It merely attests to the fact that those forms of life that are most responsive to a particular environment tend to survive better there. It does not imply that those that do are themselves qualitatively better. If a nuclear war were to devastate the planet, it has been speculated that the only survivors would be cockroaches, a very durable species. Yet this would not make the cockroach better than a human being, only in certain critical respects better acclimated to that dramatically transformed and fundamentally shattered world.

The dinosaurs were, by and large, extraordinarily well adapted to their own Jurassic and Cretaceous surroundings, remarkably successful in their day. Yet when those supportive environments changed, arguably as a result of a devastating comet or asteroid impact, many of their previous assets, such as their great size, became genuine liabilities. The meaning and value of their adaptive strategies was dramatically altered as their contextual landscape morphed. It revealed that environment to be, rather than a stable reference point for what was "better" or "best," no more than perennially shifting sand. For mankind in particular, enduring meaning must derive from the, as yet, dimly-perceived spiritual realm to which we are individually and collectively evolving, and in relation to which the rules of the game, though different in emphasis from those of the Darwinian struggle, as classically conceived, are not nonetheless fated to remain a permanent mystery.

It has been hypothesized that the drive to colonize space is, essentially, for humanity, a "run for life" a race against time to seed other worlds with our representative progeny before we unconscionably, yet—as it would appear—inevitably, destroy our own troubled planet. A 2010 Guardian article articulates this idea, as put forth by Stephen Hawking:

> The human race must colonise space within the next two centuries or it will become extinct, Stephen Hawking warned today. The renowned astrophysicist said he fears mankind is in great danger and its future "must be in space" if it is to survive. In an interview with website Big Think he said threats to the existence of the human race such as the 1963 Cuban missile crisis are likely to increase in the future and plans to handle them must be put in place now (Press Association).

But this argument contains an essential fallacy, an inconsistency in time among its component elements. For the historical splitting of the atom occurred well before any corresponding ability to transport a human population to any known world other than the one we inhabit.

Even now, beyond the close of the twentieth century, we have yet to establish colonies on the moon or on Mars, the very closest candidate worlds, while, in the meantime, here on earth, atomic weapons have long been superseded by their vastly more destructive thermonuclear equivalents. As a clincher to this argument, it can be readily assumed that the same technologies which could transport us to the Moon, to Mars, or to any other conceivable world, would be able to carry our nuclear-tipped weaponry there as well, and far more easily than it could any formerly earthbound population. The tragic result of such an apocalyptic launch would be to erase in a single moment of wanton destructiveness what the strenuous efforts of the decades and the hopeful ponderings of the ages had hitherto been able to achieve.

This space-race-escape argument is furthermore, not only specious, but is, like an ideological Maginot line,[1] hazardous in its perfidious impact on our contemporary will and thought, for it would purport to relieve us of the unavoidable responsibility to change as we certainly must. Running away from our problems can be no more effective as a collective solution than as an individual one. Our species trial is upon us, and it can no more be avoided than its existence can be successfully denied. The only reassurance that this otherwise misleading argument can give is one having little to do with us and with our personal hopes for the future, for it may form a plausible basis for a belief in benevolent versus malevolent aliens. This would logically seem to pertain, as the members of any other advanced species must once too, like us now, have confronted a comparable peril and found the strength and compassion to live in planetary harmony with one another, before earning the subsequent right to reach for the stars. What remains for us at our own historical and evolutionary crossroads, and during this unimaginably critical period of time, is to become either like these envied aliens in our own future glory— exploring space and giving birth to wonders as yet unimagined—or to become like the dead dinosaurs of our own extinct past, concluding our wayward trek through historical time only by fertilizing the soil of the future, near or still fairly near, with our disconsolate remains. This favorable or unfavorable end will be determined, in the final analysis, by how precariously our knowledge has already outpaced our wisdom, and on our resulting ethical and moral maturity.

Those who have, in lieu of spiritual ideals, misappropriated Darwin's theories to create coarse and competitive systems of value for human beings based on animalistic norms have done mankind a terrible disservice. In the case of Nazism, based on Hitler's view of a struggle between races, as a twisted, virulent, and particularly unscientific form of social Darwinism, they have led to colossal violence and destruction.

Such a crass glorification of material conflict, in the absence of spiritual values, is commonly acknowledged to end for modern man in a tragic, bloody, and dismal orgy of collective death. Yet, in a far more subtle way, a more credible form of more limited Darwinian ethic still informs the commonplace tenets of modern life, a life in which profits are routinely valued over people, and people themselves measured in dehumanizing relation to material norms alone. This is the type of world we live in, one in which, though not utterly consumed by violence, in accordance with the Hitlerian motif, we often find it burdensome just to wake up in the morning and to head out our front doors, for it has become a treacherous domain of needlessly enhanced suffering, friction, and tragically-obscured meaning.

We can use the analogy of a carload of kids squabbling on a long trip to exemplify this point. Each child, being impatient and restless, seeks to promote his individual happiness, by taking his frustration out on the others, making it an unpleasant ride for everyone. The driver of the car is, moreover, distracted by their disruptive antics. This increases the risk that this back-seat unruliness will lead to an unfortunate accident from which all will inevitably suffer. Whatever temporary advantage any manages to secure, the most conspicuous general outcome is a diminished total environment of enjoyment and happiness for all concerned.

We should get the sense, however vague, that we are on the wrong track when we see how, when we behave in a selfish and animalistic way, we are not happy, either individually or collectively. Acting in full accord with the Darwinian ethic, we become a burden to ourselves and to others, and, in so being, suffer under the influence of emotions that Buddhists term "afflicted." Like the tragic couple in the legendary Garden of Eden, in choosing the animalist way, the road of our material nature alone, we take the dark and violent path away from our best and proper destiny.

That the way of the brutes is not the way to either happiness or evolutionary progress for mankind is indicated by an abundance of real-world examples. The mafia chieftain, dining at his own restaurant, must look over his shoulder between bites, for he never knows when those whom he has brutalized may seek to return the favor. By the same token, First World nations which have, in the extreme, made warfare and terror cornerstones of their rule have typically self-destructed. The Nazi state, based on militarism and on a functional modality of violence, carried war beyond the accomplishment of its practical aims to precipitate the nation's ruin.

The Nazi experience has been referred to in this book repeatedly as the premier example of how high technology, when dangerously conjoined to low morals, produces, in the modern era, a formula for obliterating the world. As a further example, the system of Stalinist terror, one that could likewise be sustained only by further terror, stifled the creativity of the Soviet people, thus ultimately undermining that society's ability to compete. In an anarchic world environment, in an era of advanced technology, where each seeks security at the expense of the rest, all together are insecure; where some form of cooperative world order exists, true peace and security are possible.

A subtle problem today for many advanced nations, the United States among them, is that the dominant mode of religion is strictly faith based, with its values rooted in mystery contributing little scientifically to a pragmatic theory of compassionate action. Its ritual-based nature, and characteristic emphasis on decorous rites alone, however valuable and enriching such rites may be, have proved insufficient to bridge the gap between reason and faith, let alone between faith and science. Nor, in their current arcane and suppositional form, can we expect such methods to have enough of an impact on the often cruel, indifferent, and thoughtlessly competitive norms which govern our daily behavior to outweigh the fully-verifiable claims of a partial-ly-completed science. Revisiting the analogy of the kids in the car, our collective journey through life has thus become, and so, precariously remains, a distressing and potentially dangerous one.

Many who lovingly kiss their families goodbye on their way to work, are as just as likely when they arrive there to undermine by treachery someone else's ability to feed *his* family. Within the average corporate environment, dog-eat-dog norms are notably in ascendance. So much more cogent than faith is the influence of an incomplete, yet, to that extent, verifiable science in its practical influence on our everyday actions that most of us feel compelled to behave as if we were personally engaged in the classical aspects of the Darwinian struggle in each and every moment of our lives. We think, despite our hopes, qualms, instinctive compassion, or intuitive inklings to the contrary, that we have to behave this way, because that is how the world itself operates. In the modern era, we are desperately in need of a method that is both practical and palatable to rational minds of cogently demonstrating otherwise, and of revealing the higher dimension of spiritual laws genuinely applicable to mankind, in a world now so critically in need of such knowledge.

The characteristic Eastern approach to religious understanding, common to Hinduism, Buddhism, and Taoism, in being both spiritual and experiential, posits a higher way of knowing in which moral truths can be verified, though by methods substantially distinct from those of a more customary subject/object empiricism. It offers the intriguing possibility of making religious values scientific in their own right, and, in the process, gives us hope that we may finally bridge the gap between an all-too-materialistic Western science, and an-all-too-belief-ridden Western faith, which, in the modern era of dazzling scientific breakthroughs, lies moribund on the pyre of mystery. The former faith may thus prove, in its own unobtrusive way, to be the unlikely salvation of the latter, though the latter, in defense of its own traditional trappings, trappings which cannot nonetheless obscure a common substance, is likely to come to terms with it at best reluctantly. It would seem that, in the modern integrative world of relentless technological advances, it is only by embracing its commonality and inherent oneness with other leading faiths, that Christianity can adequately defend itself against the onslaughts of a "scientific" materialism whose specious rationality is nonetheless more of a hangover from the nineteenth century than an expression of understanding in the ultra-modern age.

To summarize, the integrative perspective of the mystic, parallel in certain key respects to the evaluative method of the quantum scientist, offers the intriguing potential of substantiating through personal experience, rather than through blind faith alone, the evolution of mankind toward finer energies and more benevolent interactions on our collective journey to a more enlightened world. It contains the potential to resolve the age-old epic clash between science and faith, one which traditional religion itself, overly-reliant as it remains on faith-based methods and static doctrines alone, has been historically ill-equipped to resolve. Interestingly enough, science, as much as faith, seems destined to be swept up into this integrative merging and to a notable extent already has. As Lipton and Bhaerman describe:

> We are now speeding toward our third transit of the balance point between the spiritual realm and the material realm. What lies before us when we arrive will be defined by our choice between two alternative paths. We may choose to stay in the same familiar world of dueling dualities, wherein religious fundamentalists and reductionist scientists continue to polarize the public. This path will obviously continue to take us toward the same destination we are heading to now— imminent extinction.
>
> Or, as we return again to the balance point, we may choose to resolve our differences by seeking harmony over polarity. By combining formerly factious elements into a unified functional whole, we can open the door, transcend historic dualities, and experience an evolution that will provide for a higher-functioning more sustainable version of humanity (207).

NOTE

1. As concisely described in Webster's New Explorer Desk Encyclopedia, the Maginot Line was an "elaborate defensive concrete barrier along the German frontier in NE France built in 1929-34. Named after its principle creator, Andre' Maginot (1877-1932), it was supplied with heavy guns and had living quarters, supply storehouses, and underground rail lines. However, it ended at the French-Belgian frontier, which German forces crossed in May 1940" (738). Its larger and symbolic significance concerns the fact that it gave the French a false sense of security which restrained them from taking the substantive steps needed steps to prepare for the Nazi invasion.

Chapter Eight

How We Perceive the World

When the Buddhists say that our suffering is rooted in ignorance, what is it they claim we are essentially ignorant of? It is two things primarily: 1) the fact that change in the world of phenomena is continuous, and 2) that all we can separately identify, including our own physical natures, is in fact interrelated to other elements of our environment. It arises from a concatenation of causes, and is part of a larger whole. Change one cause and you will alter the effect of what is perceived and identified as well.

Here our mistaken approach to reality is not a conceptual ignorance only, but a routine flaw in the way we respond to everyday situations, to other people, and to life. The deeper purpose of Buddhist meditation, so defined, and as just one such technique, is to bring us face to face with the nature of life at the root level, so that in relating to events accurately, we may radically minimize our suffering. That we enjoy and experience life is not the problem; it is the fact that we relate to it wrongly, by perceiving it inaccurately, that causes us the greatest agony.

With respect to continuous change, let's say that, in the midst of my daily routine, I tell my spouse, I am going to "my favorite restaurant" to get my "usual cup of coffee." Here at least five separate items are identified: self, spouse, restaurant, cup, and coffee. All are presumed to have retained enough of their distinguishing characteristics since the last time I identified them for me to refer to them again understandably. But let's examine their reality more finely.

With regard to my favorite beverage, although the cup of coffee I order "again" at my favorite restaurant may look, feel and taste the same as the last one I imbibed, it is a different offering entirely. This latest cup is not the same as the original, though it may have the same standard appearance and dimensions. The previous container has already been thrown out and perhaps already been recycled. The coffee that I enjoy so much with its two cream, two sugar combination is also likely to taste very much the same as the original beverage. Yet, that previous coffee too has already been consumed and digested, its liquid absorbed or excreted. The new beverage has an admittedly very similar composition to the one previously consumed. In fact, both the restaurant and I have aimed to make it so, in order to duplicate and standardize the desired experience. And although I still have a tongue, it is not exactly the same tongue that I had when I tasted this presumably identical drink for the first time some years ago. The cells in my tongue have regenerated, and if I look at a picture of myself from an earlier enough time, I will likely note, not just the continuity, but the change in my existence, as well. The restaurant I go to for my coffee will also have subtly changed from the last time I went there. It could now be under new ownership; the attendant who hands me my similarly-constituted cup of coffee may be a different one as well, though I am not as likely to notice these subtle transformations in the restaurant environment, as chairs are moved and table-tent signs replaced. Yet, should I move to another town, only to return to that restaurant some 15 years later, it might not even be there anymore. This may cause me to feel, in some strange and unsettling way, that part of me has vanished or died as well. This is, in fact, and in a very real sense, exactly what has happened—at least on a purely physical level—and as the evidence of our own senses reveals. We witness numerous deaths within life, in the transition from what was to what is. We simply maintain enough continuity in the tapestry of our experiences to hide this fact from ourselves.

Now when I finally return home from my trip to the coffee shop, my wife who was pleasant when I left, is found to be consumed with anger. I got *myself* a cup of coffee, but did not get her one. So now I have, at least for the moment, an angry wife instead of a pleasant one. This is another element of change—and one that is clearly unwelcome. Within the context of this hypothetical scenario, restaurant, self, spouse, cup, and coffee are all revolving parts of a transformational matrix in which no constituent element ever remains exactly the same. Now let's focus on the second noteworthy aspect, that of their interrelated identity.

While the coffee has pleased my taste buds, my wife's anger has negated the bliss of imbibing it; so now I have decided to mend fences by taking both of us to go see our favorite movie, *Casablanca*, which is playing in a revival of the classics at my local cinema. Because I have "seen the film before," I am confident that I already know in advance how it all will begin and end. Yet, it is truly a different set of photons that passes through the celluloid of a different film copy of the original master image, projected onto a different screen that makes its way to my visual cortex. The film itself may have deteriorated with time and been restored. The copy I last saw may not even exist anymore. But this is no different, in fact, from what we have already said about the coffee. In this case, however, as I myself am likely to have an altered mindset from the first time I saw *Casablanca*, the film, itself, may leave a different impression, a contrast now noticeable even to myself—not better or worse necessarily, only different. I still like *Casablanca*, but maybe I paid a trip to Paris since the last time I saw it, so the line "we will always have Paris" may now affect me more deeply. The interaction between my mind and senses, and what is projected onto the screen, is, in fact, what produces the perceptible event for me—not something "out there" alone— and since some of this scenario's interdependent elements have indeed changed, the overall experience is noticeably different also.

The old Zen saw that a tree, falling unheard in a forest, produces no sound, but only sound waves, is based on the realization that the experience of sound involves a coming together of experience and experiencer, both of which are in a continuous state of flux. The experience itself exists, to the extent that one may apprehend it empirically, merely as an interaction, and you can never cling to an interaction. To elaborate further, my dog could stand in for me in appreciating the experience of the falling tree (since he did not want to see *Casablanca*). In his case, the sound is likely to be richer, the olfactory component of wet wood crumbling more pronounced and satisfying. Yet the rich color would be missing, as he will see it all in what (from a human standpoint) would seem like a "color blind" perspective, and with notably less analytical reflection. If someone is drunk or high, he may perceive some of the characters in *Casablanca* walk off the screen and into the audience, or see the falling tree do a dance and run away. In that event, we would say, with a notable degree of assurance, that his perceptions have been deranged. But the appearance of the hallucination is sufficient evidence in itself to suggest that the perceptual material we are dealing with, in each and every case, is merely that of our own minds and senses, which is all that we happen to know about what we are experiencing through them. By sensory means alone, we can never see an event itself as an independent absolute. Physical eyes alone cannot pierce to the inside of Einstein's watch.

Yet whatever may be their inherent deficiencies, it is not the enjoyment of movies or coffee that's the problem. That's not a problem at all. The problem is that we refuse to recognize such experiences as they actually are. We reify them by reading a permanence into them that they do not inherently possess, and into those physical faculties that form the basis for their enjoyment. So, whenever any of these revolving elements changes in a way that defies our hopes or expectations, we suffer. And we reify them to begin with because we are afraid that the pleasure they bring is, in fact, what it genuinely is— transitory and fully unreliable. We do this, in turn, because we are not aware of any other way or of any other greater happiness. And so we insist on the verity of our fondest illusions by refusing to accept their underlying reality as changing and "un-clingable." From this standpoint, our accompanying reactions of attachment and aversion are as futile and stupid as those of the family cat who swipes at the images flickering across the screen of the living room TV.

While the reality of the moment may not seem just (with the law of karma going far toward suggesting it is), it *is* at least factual, rendering our standard responses, our catlike swipes against the three dimensional screen of life, no less futile and stupid (or as expressed more charitably by the Buddhists—ignorant). Why we do this, what we are most afraid of deep down, is what life would be like if what we currently depended on for satisfaction were absent. We liked something the first time; so we imagine and hope that it will always be there, that it will always be the same, and will always produce an identical level of satisfaction, and that for us, in particular, it *must* do so, else we will childishly punish the world (or more likely, in fact, ourselves) by choosing to be angry and petulant.

The great irony is that if we *did* let go of our fondest delusions, our minds would be clear and unburdened. We would live a much richer life, even from the standpoint of freshly appreciating those cherished objects of affection and enjoyment that we so desperately fear to lose. We would, as well, perhaps for the first time ever, be appropriately aligned with the truth—itself a propitious start on the road to happiness. We would be in a position, moreover, to at least vaguely perceive the existence of, and to entertain the wondrous possibilities associated with, the deeper, subtler and more intensive reality behind Maya's projected movie image. The catch is, we will only be able to do this effectively once we have loosened our grip on our precariously-pasted-together and forever-crumbling versions of reality, our own hackneyed storylines. Until then, since our perception of that resulting state—what the Buddhists term Nirvana—will also remain enveloped in the veil of our illusions, we are not likely to have enough faith in its unperceived reality to sustain us. So, we remain busy convincing ourselves that the impermanent is permanent, the unreal is real, and that the picture of the cake in the magazine tastes really good.

A different analogy accentuates the dangers associated this flawed, yet typical, way of life. If we are driving blindfolded on a straightaway, we will not run into any problem, so long as the road itself does not bend, and so long as our hands remain steadily and securely on the wheel. But if the road *does* bend, we will drive off the edge, most likely collide with an unseen object, and be injured. We would, of course, never consider violating the rules of the road this way; yet we do something very similar, in our routine contact with reality, whenever we blind ourselves to the true nature of our experiences. Then, whenever life itself takes a sudden turn, whenever we cling to it as needing to remain a certain way, we suffer. We suffer, as well, from changes within ourselves.

Perpetual transformation permeates our physical reality, down to the very marrow of our bones. It can be further argued, to accentuate this point, that anything occupying the sphere of change we call time, along with that which is manifestly changing, partakes in that change as well. It has to be changing too, or it would have been left somewhere behind as the clock ticked forward. If we are on the stage, and the props are being moved around us, though we may not seem to be moving with them, we are integral parts of that larger, all-encompassing drama in which ceaseless transformation is the norm.

The delusive clinging, noted above, takes place at several levels, and at some more obviously than others. The first is with regard to the objects we surround ourselves with. This is where we come to talk possessively, and with a false—yet firm—assuredness, about "my yard," "my TV," "my car," or "my cup." By claiming ownership over the objects with which we are associated, we deny the simple fact that all of the above "possessions," including our own bodies, are subject to continuous change rather than to our whims, and that our association with them may end at any time. They manifest and decay in accordance with set laws, rather than in relation to our arbitrary preferences. The more we engage in the delusive habits of control and possessiveness, the more external objects we cling to, the more we increase the odds that something will not go our way, that reality will turn the corner, and that we will suffer once again. Yet, at an even more subtle level, we guarantee our continual suffering and the painful contraction away from life whenever we create a "self," an ego, out of our own "internal furniture," out of all that is subject to change within our minds. There may indeed be something more in there, whether we call it Self or Soul, or Buddha-nature, but we are not inclined to see it through all the dust kicked up as we perpetually rearrange all of this internal ornamentation, exhaustedly, fearfully, and over and over again. We may see ourselves, for example, as pretty or smart, and invest the bulk of our happiness in those ideas, and then something happens to directly contradict them. Someone else wins the beauty contest or the quiz show.

While such events are never welcome, we suffer most through them, not as a result of their simply occurring, but as a result of our delusive attitudes. When we cling and where are caught is where we find ourselves—collectively as much as individually—at the crossroads. We can realize at such times that we are in fact clinging, that the clinging itself is misguided, and then just learn to let go, lightly and flexibly working with the elements of transformation skillfully. We can do this by first accepting their lawful nature and factual reality. The opposite and negative response is to petulantly deny the validity of adverse happenings—if not their very existence—to lash out and rage against life in its presumed unfairness, to attempt to elevate ourselves by diminishing others, to be jealous of those who appear to be benefiting from that which we comparatively lack, or to resort to an ever-abundant range of alternative, unskillful responses. Worst of all is when we choose to press the accelerator of denial through an intensified hatred, a hatred that does not forget, that self-inflicts suffering, and through which we relive our comfortably familiar agonies long after they are in other ways gone.

The simple fact that, as sensitive human beings, we are made to suffer through such negative states continually, would seem but horrible cruelty on the part of a whimsical or indifferent universe were it not for the self-concealed truth that none of what we have ever lost in the physical domain—none of it at all—has ever been permanent to begin with. Not one bit of it can satisfy us or please us forever, and most of it not even for long. The very sources of our joy, should we fear their loss or end, can easily turn to torment. To the extent that you find your house, lover, family, wealth, etc. as an indispensable source of unchanging satisfaction is the extent to which you will suffer when they manifest their protean capacity to morph and to displease. We are, moreover, not only at risk of being devastated by their absence, but tormented in the meantime by fear of their loss. When we make the possession of life rather than the love of it the focus, we always end up anxious, exhausted, jaded and dissatisfied.

In Buddhism, clinging ignorance—along with the stupidity that underlies it—is recognized as being particularly pronounced in the animal realm. As the clinging there is stronger, the territoriality, violence, agitation, and fear, are also more pronounced than they are for humankind. Moreover, to a degree corresponding to this clinging, territoriality, and possessiveness, frustration, suffering, and conflict inevitably follow. In short, the animal state, as the result of its modus operendi, as contrasted with that better world which sapient Man might choose to create for himself, is an environment marked by perpetual violence, scarcity and fear. Just as noteworthy, however, is the fact that such conditions also pertain in those environments of conflict and under conditions of war, where human beings live down to their lowest natures rather than up to their highest potential.

If we have many excessive interests and possessions upon which our happiness depends, we not only suffer to the extent that we rely on them, but accelerate our degenerative descent into suffering by repetitively battling others amidst anger, hatred, and fear in our effort to defend them. What we are called upon to do, what the human challenge consists of, is for us to see life as it is, to love it for what it offers unbidden, and to work flexibly and wisely with it as we strive onward toward greater happiness.

In the realm of international relations, we are making this same type of choice collectively. We either open our minds and hearts to change, or respond with greed and mindless clinging. We show acceptance, love and openness, or else aversion, anger and hurt. We are either welcoming the future with open arms, or clinging to the past's departed glories, resisting—

and, in the extreme, violently repulsing—the reality we collective refuse to acknowledge. We are reactively defending our preconceived views, or are open to fresh ideas. We are either behaving in a territorial way, or merging our nationalism with a progressive global idealism. We are either collectively sustaining old grudges (the Hatfields vs. the McCoys as the Israelis vs. the Palestinians) in which each side can always, monotonously, endlessly, and as it would seem to each, justifiably, draw on a lavish fund of vanished slights; or we are moving to transcend a difficult past after wisely absorbing its lessons. It is much the same dynamic at work here as exists on the individual level.

From an empirical point of view, each of our senses tells us something different about the world we inhabit. It is much like the story of the blind men and the elephant, in which each confuses the whole for the part with which he is acquainted. Here each sense is itself like one of the individual blind men, blind in this case to the routine functioning of the rest. [1]

All our material measurements are similarly relative. A distant star, for example, will appear to the senses of an observer here on earth as a tiny speck of light. Yet for someone close upon it, it is a blinding, searing, lethal, multi-million-degree mass. Nor is this discrepancy a contradiction, for it is truly and verifiably both one and the other; the difference resides with us, in our relationship to the world we inhabit.

The fact that each of our senses tells us something distinct about our reality, dividing, sifting and relatively measuring, makes each sensory modality merely analogous, in a way, to what is said to exist "out there." Like the tree in the forest that produces no sound, (but, in truth, only sound waves), all each of us, in fact, knows through our senses is our own, individual nervous system (Easwaran 35); Heisenberg's principle revisited.

This essential realization underlies what Charles Evans-Wentz describes as the "yoga of the great symbol" in his *Tibetan Yoga and Secret Doctrines.* Here the world as we perceive it is but a symbol of the higher dimensional reality, the reality of "mind-only." Once that greater state of enlightenment is even partially realized, its substance is discovered to be so much more immediate and palpable that the world, as we have hitherto known it, appears from that point forward as but a rough-hewn shadow—a veritable symbol of itself.

Within this elevated experiential context, we realize the deficiencies of sensory evidence by recognizing and surmounting our own limitations, limitations that have existed all along, but that we are now able to acknowledge and hopefully to overcome by focusing in a different direction. We are no longer bound to apprehend the world through the flickering "shadows" projected upon the walls of Plato's cave, the sensory reflections cast within the confines our limited material minds. The sunlight of integrative consciousness exists to be discovered by as many as are individually prepared for it, and, in the contemporary era, none too soon, with the "dark sun" of nuclear knowledge already risen.

NOTE

1. "The 'Blind Men and the Elephant' tale originated in India. It is widely thought the original story originated in Hindu lore. It was translated to the English language in the 19th century as a poem by the English writer John Godfrey Saxe. A version of the story has been used in the Buddhist culture as well as the Jain and Sufi Muslim culture. In modern times, the story has become widely used in philosophy and religion classes. It is used to illustrate the need for religious tolerance. The story illustrates how people form their reality and belief system on their limited experiences. In other words, perhaps each religious faith only holds truths that make up one part of God. The story is also used to teach tolerance for other cultures. We only "see" the culture in which we are immersed.

The poem illustrates how perception is based on what a person is able to see or touch. In the story, six blind men touch an elephant. Although each man touches the same animal, his determination of the elephant is based only what he is able to perceive. The poem warns the reader that preconceived notions and perceptions can lead to misinterpretation" ("Blind Men and the Elephant").

Chapter Nine

When the Past Resists the Future

In the modern, transformational environment and with respect to our traditional beliefs, many seemingly impervious to change or to systematic reevaluation, one of several conclusions can be drawn: 1) either what we have traditionally believed is completely true as stated; 2) the traditional view is false; 3) the traditional view is partially true, yet needs to be modified; or 4) that view is true in a way different from the way in which we have hitherto understood it; this last option offers the most intriguing possibilities of all.

The Catholic Church of the past, for example, saw the heliocentric theory of Copernicus as a threat to its religious authority, as his view *appeared* to contradict the conception of Man as the center of the universe. Yet did it in fact do so? Man may still be the center of the universe, the centerpiece wonder at least of earthly creation in some more profound and significant way than by reference to planetary alignments. In the final analysis, the institution's own doctrinal inflexibility proved to be then—and would be today—far more damaging as a threat to its essential legitimacy than the fully warranted assimilation of new and elucidating ideas.

From this perspective, the ancient science of yoga serves as the source for an intriguing alternative view of Genesis, one which has the distinct advantage of being not only more plausible than the literal one, but, as a needed supplement to our contemporary understanding, of cogently challenging the very applicability of Darwinism in its traditional, unmitigated, and material vs. energetic form, to the directions of our human experience. In the process, it stands to reintroduce Man to the lost treasure of his spiritual nature contemporaneously obscured within the fog of our materialistic preoccupations and amidst the unduly frenetic pace of modern life.

The lower energy chakras, those relating to sex, survival, and power, the obsessive preoccupations of our benighted and conflicted world, are associated with our limited physical and material nature. They are, in the Eastern view, what Man needs to transcend (i.e., to govern—not notably to eradicate) by ascending to the higher levels. We would thereby reacquire the veritable Eden of our original, higher nature. The upper "heavenly" chakras correspond physically to the brain and higher organs, those reputed to have emerged last in life's evolutionary expanse, even as more prosaically defined. They also, quite tellingly, correspond directly to thoughts and behaviors typically regarded as harmonious, altruistic, and angelic.

The lower chakras, corresponding to instinctive elements which appeared earlier in our planet's evolutionary chronology, and which Man and the brutes share in common, correlate not only to the darker, more primitive, and strictly physical nature of Man, as noted above, but to the brutality, aggression, disharmony and conflict of material life itself as it has tragically unfolded throughout human history. Such impulses map directly to the ground of Darwin's natural struggle. Choosing by his thoughts and actions to darkly subsist at such a level, the individual man is effectively expelled from the Garden of Eden. He is fated to make his way through life only with the greatest dissatisfaction, pain, and affliction, living with little respite amidst virtually unmitigated competition with others similarly burdened, gloomily muddling through "by the sweat of his brow" (Genesis 3.19).

The essential game of life for mankind in the civilized phase, vs. the primitive natural phase of scarcity and material conflict detailed by Darwin, is predominantly a game of energy. Its challenges test the quality of our developed humanity, rather than our outward physical strength or survival status alone. That more rudimentary element has already been strengthened, through earlier trials of primitive struggle, to serve as a sufficiently reliable platform for a more profound and meaningful advancement, one that we are collectively experiencing now. In this latter, more developed phase, it is not the number of objects we are surrounded with, the possessions we acquire, the approval we receive, or the alliances we form, but the thoughts we habitually entertain and the choices we make routinely from moment to moment that truly count the most. And, although the proving ground for this civilized-level struggle is the domain of our social interactions, its workings are not governed by the prevailing whims of social custom. Nor is victory in accordance with them decided by any societal nod of approval, for they are as much an integral part of life's undergirding as are those natural and scientific laws with which we are more directly familiar.

Mankind has evolved from lower to higher form; the next step in the upward progression is, at basis, spiritual—in elements of growth beyond form. Those who, in the vanguard of this progression, are most prepared for it are also most likely to misfit, and to be most distressed by the old goals and hackneyed ways of mundane competitiveness. These are goals that, for them, are less worthy than the humanity that is thoughtlessly sacrificed in the process of their attainment. By the terms of this higher evolution, a choice in favor of compassion will propel us further, not just for the collective good, but for our truest individual good as well, than one motivated by the demands of physical survival or of self-preservation alone, for to choose in favor of the former is to affirm the primacy of our spiritual natures. It is to be properly aligned with the crucial goals of our current evolutionary phase. This makes Christ's willing crucifixion, an otherwise nonsensical act when viewed from the standpoint of the old, violent and competitive rules alone, in fact - not merely in hope - the most evolved act of the most fully evolved being.

Thought directs energy. It creates our specific signature. It defines who we are. On the basis of who we are, we determine our own futures in who and what we attract. Thought and energy thus establish our larger destiny. With respect to this energy, we are like bees gathering pollen, only what sticks to us are the vibrations associated with our thoughts, and as manifested through our actions and choices. Love, generosity, patience, tolerance, and courage are among the higher energies to be accumulated. Accumulate enough of them, and we will ascend higher, as surely as helium rises. Hatred, egotism, bigotry, anger, lust, and fear are the darker and heavier energies to be transmuted, as they draw us down, attracting us to the vibratory spheres corresponding to their denser quality. These latter are the environments already identified as characterized by violence, cruelty, scarcity and dissatisfaction. They are the environments of the lower realms. At each moment we stand at the crossroads, presented with a very human choice, as to whether to take the high or the low road. In terms of frequency or vibration, we must decide, in other words, what note on the clavichord to strike—high or low, narrow or open, tolerant or intolerant, with the external props on the stage of that ordinary, yet momentous decision mattering far less than what we think, what we choose, and who we consequently become.

Each of us is making these choices continuously, whether we happen to realize it or not. The arena, in which we make them, be it ever so humble, is always precisely where we are. Each of us is thus in a position to contribute, if he so wishes, and by means of a refinement of his energies, to making this world a better place, and in the very best way possible, by starting with him or herself, within the routine domain of personal thought and action.

From the physical alignment of the chakra system, the direction of humanity's evolutionary course from form to formless, from brutal to angelic, is indicated. It suggests, as inapplicable to Man, in contradistinction to the brutes, a behavior system which has its reference point in material form and in the material world alone. When the lower nature predominates, one reaches out greedily into the transitory domain of physical sensation to delusively lay claim to more and more possessions, never to be truly satisfied; as one reascends, he increasingly relinquishes more—gives more and serves more—in order to progressively manifest more of the intrinsic heavenly qualities of his higher nature, and to benefit from its markedly superior modes of enjoyment, satisfaction, and understanding.

The modus operendi of the higher chakras is thus one of giving rather than of taking, of serving rather than exploiting, and is, in its characteristic expression, the polar opposite of the traditional Darwinian imperative as it relates to the limited ends of physical survival and material acquisition alone. Yet it is precisely what is correct for mankind, and what puts us in proper alignment with the cause and effect dynamic of the higher, karmic law, the law by which we receive by offering of ourselves, thereby magnetizing through the accumulation of the more refined and higher energies associated with the quantum—or spiritual—level of activity, all that we genuine need. We are thus able to prosper by implementing a mode of life and of human relationships separate from, and superior to, the violent imperatives of the classic Darwinian struggle, with its more limited and dissatisfying set of pyrrhic rewards parsimoniously allotted within the sparse desert domain of materialism. Alternately expressed, it puts the individual man in alignment with the higher principles of manifestation whose abundant outcomes are commonly identified in the religious lore of various cultures as nothing less than "miraculous," yet which are not beyond the scope of the individual man who properly lives up to his divine potential.

As Jesus states of the individual who is fortunate enough to reclaim his soul, "don't you know that ye are all gods, and that ye are all children of the most high" (Corinthians 3.16). As Saint Augustine comparably asserted, "miracles are not contrary to nature but only contrary to what we know about nature." It is by progressing in alignment with such higher "spiritual" laws, that the will of God, so to speak, becomes manifest through mankind individually, and, by ultimate extension, collectively, forging amidst the unfolding of their unobtrusive yet ineluctable workings, the harmonious world we have sought in vain elsewhere throughout the pages of our torturous history. As Swami Satyananda Saraswati notes with regard to Adam and Eve:

> Their "fall" occurred when they lost contact with the deeper core of their being, when they became selfish and totally lost in the world of objects. The Garden of Eden represents the world. It can either be experienced as a heaven or hell; it depends entirely on the level of harmony, understanding and awareness. Before Adam and Eve ate the first fruit of the tree of good and evil, they were in the heavenly state, in the higher chakras or sahasrara; after taking the forbidden fruit, they descended to the level of mundane experience and perception. The Garden of Eden did not change— the world remained the same— but their relationship, their understanding and their identification completely changed. And this story is trying to tell each and every person something important: that the world can be either a place of bliss or a place of dissatisfaction depending entirely on your level of awareness. Either one lives at the insipid, semi-conscious and joyless level of the mooladhara chakra, or one

ascends to the exalted, meaningful and blissful level of the higher chakras by harmonizing the mind. This is the way to transform a veritable hell on earth to a heaven on earth (553).

What we have thus inherited from Adam and Eve, the "original sin" of our Catechism classes, is, at basis, the inheritance of a defective consciousness which, as expressed by humanity at a level beneath our proper nature, conduces collectively to a world, not only brutal as with that of primitive nature, but genuinely demonic in the protracted scope, and unnatural domain of its attendant destructiveness. How sad, yet interesting, to reflect that such a basic, yet critical, error, committed in a figurative garden of bliss long forgotten, a descent from a paradise of higher virtue and functioning, is now the very thing, the crucial paradox and evolutionary inconsistency (with knowledge in service to ignorance) that has planted us at the doorstep of collective self-annihilation.

In the parallel view of Buddhism, though mankind and the brutes occupy the same terrestrial space, the animals, along with the so-called hungry ghosts and hell beings, are denizens of an experiential domain which encompasses the so-called unfortunate realms. The gods and humans reside in a relatively favored ground of experience. They, as well, are nonetheless subject to the inevitable dissatisfaction issuing from the loss in time of these relatively favored states.

Yet if there *are* higher and alternative realms to those to which our physical senses are tied, then why don't most of us sense them? The Eastern answer is clear, as indicated above. Our minds have descended and become fixated on the lower levels associated with materiality, so our experience has coalesced there as well. The challenge for the individual man is to ascend through the chakras to a level of unitive or integrative consciousness, where our higher faculties are destined to reveal what is currently accepted only on the basis of faith, reasoned hope, or incipient feeling. We can, as such a knowledge is progressively realized, be expected in our national, social, religious, economic, and other concomitant relationships to manifest in the outward world the newly benign perspective associated with this heightened view. Nor would it be necessary for all human beings to reach this state for its benefits to be collectively realized.

To use an appropriate analogy, once a submarine reaches a certain depth, its hull will predictably burst. It need not descend completely for that outcome to be thoroughly manifested. Likewise, because the consciousness of each individual is, at a deeper level, conjoined to the collective consciousness of humanity, whenever one person advances, he cannot help but raise the whole of which he is a part to that correspondingly higher level. Also, per the chakra system, with the crown chakra dominant, we can expect an enlightened individual functioning from that plane of higher activity and understanding to exert a greater, and even preponderant influence on the world no matter where he resides, giving modern meaning to the Old Testament adage that if God found ten honest men he would save the city. In modern times, by extension, once the advancement of even a portion of our common humanity occurs at that deeper level of consciousness, it may represent the hairs-breadth difference separating planetary survival from the extinction of all life on earth.

In this connection, Wayne Dyer in his book, *The Power of Intention*, summarizes David Hawkins' estimate of the impact a single individual's spiritual advancement would likely have on the collective well-being of humanity:

> One individual who lives and vibrates to the energy of optimism and a willingness to be non-judgmental of others will counterbalance the negativity of 90,000 individuals who calibrate at the lower weakening levels.
>
> One individual who lives and vibrates to the energy of pure love and reverence for all of life will counterbalance the negativity of 750,000 individuals who calibrate at the lower weakening levels.
>
> One individual who lives and vibrates to the energy of illumination, bliss, and infinite peace will counterbalance the negativity of 10 million people who calibrate at the lower weakening levels (approximately 22 such sages are alive today).
>
> One individual who lives and vibrates to the energy of grace, pure spirit beyond the body, in a world of nonduality or complete oneness, will counterbalance the negativity of 70 million people who calibrate at the lower weakening levels (approximately 10 such sages are alive today).
>
> One single avatar at the highest level of consciousness in this period of history to whom the title *Lord* is appropriate, such as Lord Krishna, Lord Buddha, and Lord Jesus Christ, would counterbalance the collective negativity *of all of mankind* in today's world.
>
> The negativity of the entire human population would self-destruct were it not for the counteracting effects of these higher energy fields (106-7).

Where all people in a group are essentially alike in nature, the subtle "existential" influence of any one individual on another, or on the group as a whole, is usually not discerned. But place a preponderantly spiritual individual in a materialistic crowd or visa versa, and the effect of the more notable difference is often immediate and palpable. Like bathers sharing a common pool, we are together immersed, at the primary or quantum level, in a placid or ruffled reality established that one way or the other by our mutual influences and by the sum total of our collective thoughts. This gossamer inter-weaving, in turn, forms the fabric for the peaceful or violent world we must ultimately, collectively—and lawfully—inhabit. Those who are at peace real-ize peace for all, and, whatever their outward role, do so most effectively and enduringly, and more so, by far, than those who would have us advance along the road to progress through violence, hoping by dark and disruptive means to reach the alternative end of light. In espousing violence, even for an ostensibly altruistic purpose, they beckon and ignite a compensatory vio-lence, and whether the pendulum of doom returns to greet the perpetrator sooner or later, return it must as a matter of course and inevitably.

Self-realization, with the benefit of Life in mind, is the characteristic method of the quantum or spiritual revolutionary. It is one which, in integra-tive fashion, seamlessly bridges the divide between those who seek man-kind's material benefit (the clothes on our backs, the shoes on our feet, and the food in our stomachs, without which attention to a deeper happiness is effectively barred) and those who look to the requisites of our ultimate spiri-tual destiny primarily. The integrative state beautifully blends the one with the other.

The individual, functioning at the quantum, or spiritual level, in acknowl-edging the limitations of his separate intellect, however well-informed or enlightened, does not attempt to substitute his blueprint for Life's, or his opinions for those of others. He seeks instead to harmonize himself with the higher will of what is ultimately recognized as an orderly and intelligent universe. He does this for the purpose of best realizing, both his individual good, and his society's collective betterment; his approach, were it to be described in conventional terms, would be "not my will but Thy Will be done."[1] As consistently stated in the language of the *Tao Te Ching*, "As for those who would take the whole world / to tinker it as they see fit / I observe that they never succeed; / for the world is a sacred vessel / Not to be altered by man./ The tinker will spoil it; /Usurpers shall lose it" (Tzu 95). The truly spiritual (rather than the merely religious or doctrinaire) individual, whatever his ostensible denomination or outward material role, positively impacts the world as much or more by who he is as by what he does, with his actions proceeding inexorably from and along the same lines as his nature.

The way in which the opposing process of material revolution degener-ates is by no means invisible either, with the historical effects of violence, abuse, and neglect traceable in the legacy of our past collective actions and in their continuing impact on our present experience. As much has been said on this already, with more to be further elaborated, we will not belabor the point now, except to trace this process in its typical and more general outlines. By initiating conflict, we effectively create our own opposition; we define it (i.e., set its identifiable limits within the bounds of our perceived reality. We conjure it forth as something distinct and inimical to us). In one stroke of a

prejudicial brush, we foster our "enemy's" self-identity, while simultaneously congealing the basis for his enduring rancor; the more we brutally oppose
him, the more we compel him to refine and strengthen himself, until, in
Ahab-like fashion, fully-armed and ominously effective he plots and exacts
his revenge. This is the typical course and fate of the merely material revolutionary, whether proceeding from a leftist or rightist imperative. Theirs is
characteristically a political agenda rammed down the throats of those who
neither want it, nor are prepared for it, with history replete with examples of
how such an approach must and will inevitably fail by consuming and
contradicting itself.

The *Tao Te Ching* outlines the more subtle, effective, and enduring approach of the quantum revolutionary in stating that "High virtue is at rest; / It
knows no need to act. Low virtue is a busyness / Pretending to accomplishment" (Tzu 106). This is not, as it might appear on the surface to be, but a
cavalier prescription for laziness, indifference, or neglect in either the need
for the individual to fulfill his purpose or for the world itself to change. Nor
is it a jab against those who struggle long and hard for success or for the
betterment of their respective societies. Its message is infinitely more subtle.
The paradox here is that, although the spiritual individual sees himself as
doing nothing of himself *alone*, by foregoing the coarse and unwarranted
imposition of his separate will and thought on others, so typical of the material revolutionary, *through him* miracles of harmony and mutual accord are
achieved. Through stillness comes a subtle harmony; through harmony proceeds right action in a measure appropriate to its time and place; through
right action, the needs of society, world, and self are met.

This is consistent with Milton's conclusion in his poem, "On His Blindness" where he expresses his fear that losing his sight will fatefully inhibit
his ability to serve God and the world through his talents. His dread is that, in
"hiding his light under a bushel," he will be subjected to an unfavorable
judgment (Milton). Acknowledging the Ultimate at the poem's end as the
power behind his actions, his conclusion with reference to God (and thus to
Life) is essentially the same as the *Tao Te Ching*'s: "his state is Kingly /
Thousands at his bidding speed /, And post oe'r Land and Ocean without
rest: / They also serve who only stand and waite" (Milton). In Buddhist
terms, the Dharma (what Milton and others would refer to as God's underlying will) is inexorably unfolding, and we remain one with that creative process whether we appear at the moment to be standing still, or not.

This quintessential view is one, as well, with the prescribed approach of
the Hindu *Bhagavad Gita*, which categorically states that it is in our nature as
human beings to act, to participate in the unfolding drama of creation in some
outwardly productive and inwardly beneficial way; it challenges us to observe that even by standing still, we are always in truth doing *something*. Our
bodies are swept up in the flow of time and in the relentless processes of

material change; our blood continues to circulate; subatomic particles perpetually revolve around their nuclei within the atoms of the molecules of our cells, with those cells themselves relentlessly regenerating. Nothing in the physical world holds firm. A brutally direct appreciation of this fact reveals that the key to progress, individually as much as collectively, is to yoke the material activity of the individual to the larger processes of the Universal Life, hence the term yoga. Its outward expression is for the individual to engage in correct action by undertaking the work and assuming the role that is at once best suited to him, and to the needs of society as a whole.

Because we are one with Life anyway, beyond any boundaries we may have artificially set for ourselves, when we act in harmony with our truest natures we inevitably act for the benefit of all. We become, in effect, the living universe harmoniously providing for itself. In this humble, yet exalted capacity, rather than as one man crying in the revolutionary wilderness or with his fist held high in the tempestuous air—*unless that is what the dictates of Harmony themselves necessitate*—we most powerfully and enduringly, yet seamlessly, achieve. For who can hope to conquer the demons of a stubbornly conflicted world, with all the malevolent impetus of their entrenched dominion behind them, without the subtle force of a spiritual ocean behind him?

Marx, who labeled religion "the opiate of the masses,"[2] saw his basic theory as Darwinism applied to politics with conflict—in his case among classes rather than species—as the catalyst of social change. The way of Marxism-Leninism, in particular, has been to manipulate, through the processes of legislation if possible, but with the tools of violence if necessary, the external props on the revolutionary stage of history. It forces them to conform, amidst anger and angst, to the insurrectionary's fractious will. The way of the spiritual revolutionary is to acknowledge both the immediate and enduring value of harmony in a world where, as *The Voice of the Silence* proclaims, "The pepper plant will not give birth to roses, nor the sweet jessamine's silver star to thorn or thistle turn" (Blavatsky 34). The seed grown will thus logically and inevitably be one with the seed sown.

With respect to what we might fondly hope to achieve in the way of material betterment, as noted by the Great Sage, Babaji, and recorded in Yogananda's *Autobiography of a Yogi*, "Few mortals know that the kingdom of God includes the kingdom of mundane fulfillments. . . . The divine realm extends to the earthly; but the latter, illusory in nature, does not contain the essence of Reality" (311). Here the "projector room" for the material realm is to be localized in the domain of thought. It is there that the reel must first be changed for the conditions we observe to be outwardly transformed.

This is the view from the highest level, for the world is not illusory in the sense that events are not happening, but in that, as Heraclitus observed, we can never step into the same river twice (no matter how similar the waves on the surface may appear to be). The illusions on the screen of our individual lives are part of a reel that is set in motion at birth, one which does not stop until the day of our deaths—and perhaps not even then. This remains true, however much one moment may ostensibly resemble the next, for, as we observe the world, the observer himself is changing. As Camelot is created, we digress by the same movement toward the return of Excalibur.

This is not said to make us cynical, only to make palpable the truth that there are limits to any change that is exclusively material, and that does not at the same time transform the heart of the woman or man experiencing it. It is not the nation or other abstraction that passes through the portal of death into a higher realm of experience, giving life itself its enduring value. It is rather the essence of the individual person, hopefully strengthened and enriched by her or his material experience as the meaning behind material creation. It is our deeper, more spiritual values that must inform our material actions in order to render them meaningful, beneficial, complete, and whole.

To do justice to Marx, organized religion (as contrasted with true spirituality) *has* served, or more specifically stated, mis-served, on more than one occasion to opiate the masses by sedating them into an artificial and obscene tolerance; but that would apply only to religion in its false and hypocritical guise, rather than in its true and legitimate form. True spirituality does not cut a swathe between the spiritual and the material, between this world and the next, but is of benefit to mankind at all levels integratively. That it should appear otherwise is itself one of the great ideological myths of the Western world, one that a cross-reference to Eastern practices, such as spiritual meditation can serve to ameliorate.

Too often, in fact, has spirituality in our hair-shirted Western culture been dualistically contrasted with "the world, the flesh, and the devil," as if to be happy in the future inevitably meant being unavoidably miserable now. This has led many who are highly pragmatic to turn their backs on religious faith entirely, trailing behind them those who find it hard to dismiss from their compassionate thoughts the world's material suffering. The traditional dichotomy of spiritual and material has been humorously expressed in the joke about the two Puritans, where the one says to the other, "it sure is a beautiful day," and the other replies, "I know; we will pay for it later."

Among that which is integrated in the integrative state are the joys of this world and those of the next where, as the Tibetans proclaim, "all good things come from all good actions." This is not only true beyond the portals of death, but true in life as well and across the board. The renunciant, in actuality, renounces only that which is productive of his misery (i.e., his ignorance), spiritually cleansing and refining his senses, while simultaneously reducing the pains associated with his false and ignorant attachment to passing conditions. The materialist, by contrast, mesmerized by life's spinning wheel, tries to reach for it and to hold it still in its glittering passage, and is, in the process, tragically injured and burned. Material ownership, while a practical and legal necessity, is neither a law of physics nor an accurate expression of the way things are. Misery inevitably accrues from any delusive attempt to hold back the flow of life, to claim ownership over time, or over that which has never and can never be owned.

In the final analysis, the internal choice of selfish versus selfless may be the truest avenue to true spirituality and to universal material progress. As the subtlest of subtle truths, the narrowest of narrow gates, it challenges us to divest ourselves of our more mundane and stifling self-preoccupations. It is not just the Mercedes in the driveway, but the Mercedes in the clouds that constitutes our limited view of a purely personal heaven, one, which for many, is no more than a "happiness deferred" for what they have been unable to achieve on earth, that is at issue. This is the pseudo-goal that must be bypassed to gain entry into the bona fide realm of bliss and eternity, which is at once the source of our highest happiness, and the integrative realm of miracles and manifestation. And we do not have to be dead before the benefits of this realm are realized either by us or by the larger world. The truly spiritual person finds his or her salvation through others, through the satisfaction derived from an integrative sense of community, or in the words of the *Tao Te Ching*, "Denying self, he too is saved./ For does he not fulfillment find/ In being an unselfish man?" (Tzu 66).

Amidst any effort to promote social change, Buddhism, for its part, acknowledges that suffering is not the exclusive province of the poor. There are, in fact, those who remain fully happy with the basics, whose level of virtue and resulting quality of life are far better than those whose souls are ravaged and scarred by their unmitigated greed. Here it is greed itself, rather than any outward material circumstance, which is acknowledged as the essential affliction, with happiness posited as an attainment extending well beyond—while encompassing—our cherished collection of toys. It is a state resulting from the beneficial recalibration of our individual and collective priorities.

By reaching out into the external world for props to affirm our worth, we deny our inner sufficiency, immediately manifesting our learned inadequacy thereby. We then proceed to enlarge that self-created sense of lack into an all-pervading *weltschmertz*, an implied doctrine of scarcity which tragically obscures the good that was, is, and has essentially remained ours all along. Inwardly dark and dissatisfied as we then must be, we project our cultivated neediness violently and catastrophically into the outside world. Such is the very taproot of war. The tragic irony is that we are now constantly seeking, through a modification of protean and external conditions, a happiness, satisfaction, security and independence which the original seeking, itself, had obscured, and which had been our effective birthright from the very beginning.

Historically, the pattern of the spiritual revolutionary has been exemplified, at its highest level in the traditional Mahayana Buddhist ideal of the Bodhisattva as an individual of service. Having earned the right to enter Nirvana, the equivalent, one might say, of the Christian's heaven, he pauses at its threshold, as he cannot personally endure the cries of pain emanating from those who continue to suffer. He, as a result, and as Buddhist lore describes, postpones his own salvation, reputedly for eons to come, so that others, even those beings occupying worlds of hellish torment—and perhaps they most particularly, in being the ones that are suffering most—can be rescued.

This methodology has its precise social equivalent in the stance of those activists, who, during the civil rights struggles of the 1960s, left their safe and comfortable suburban homes, to fight for racial justice, only to be bled and pummeled in that sacred battle. It is an approach which stands in stark contrast to the attitude of those who are fully content with the thought that they are among the few "saved" (by a God who presumably loves all) while fully believing—and blithely accepting—that 90% of their fellow beings–those they have known and loved throughout their lives here on earth—are predestined to burn in hell for eternity. There can be no greater denial of community, and of the integrative (i.e., genuinely spiritual) ties that bind, than such commonplace "holy" indifference.

Here a conflicted form of religion, one which permanently divides humanity into "the saved" vs. "the damned" has its tidy counterpart in an equally divisive politics, whose adherents are characteristically content, without qualm or sense of moral contradiction, to let the masses "eat cake," while promoting an isolated betterment, duping the society as a whole into believing what is good for the advantaged few is invariably best for all. In the corrective approach of Buddhism, "salvation" to the extent that it is personally realized is seen not so much of an individual as a universal attainment. In fact, to seek enlightenment for oneself alone is, in the Mahayana view, almost a contradiction, in that it is selfishness (along with the ignorance that forms its base) which suffuses all those issues of collective dissatisfaction and violence which continue to plague humanity.

Most of us, as children, played a game called "telephone." One child would whisper a message into the next one's ear, to be whispered in turn to the child after him, and onward in succession. The message reported back by the last child in the chain is typically found to be amusingly different from the original transmission. Revealed religions have been playing "telephone" for centuries, long before the invention of that specific device. Their messages have been forwarded through abbots and mullahs, popes and ayatollahs. Some of these esteemed religious representatives, unlike the innocent participants in the children's game, have had personal or political agendas.

Such agendas may well have induced—at least some of them—to do otherwise than to pass along their assigned messages accurately. Both Christian and Islamic institutions have simultaneously played potentially conflicting secular and religious roles. In the modern world, in particular, we now need access to spiritual experience in a more direct and palpable form as a supplement to doctrine and as a touchstone for truth to keep us reliably on course.

There is likewise the less insidious matter of traditional meaning that has simply been obscured through the passage of time. For example, the Bible contains the statement that "it is easier for a camel to pass through the eye of the needle than for a rich man to get to heaven" (Matthew 19.24; Mark 10.25; Luke 18.25). As moderns, we are likely to envision a large dromedary attempting to sail through a microscopic pinhead, and to view that goal as impossible, impossible therefore as well for the rich man to get to heaven. But such was not the original meaning at all.

In contemporary terms, the eye of the needle would be less like a pinhead and more like a toll booth. It referred, in its time, to a small door that a camel could pass through only after being unloaded, and, even then, only with some difficulty. An alternate definition would be that of a slim mountain passage. The original implication, in either case, was not that it is impossible for the camel to pass through the eye of the needle or for a rich man to get to heaven, only significantly more difficult. The rich man is distracted by his wealth, as the camel is encumbered by its bulk, in making his way to his intended destination.

We may be amply aware of the pitfalls associated with this particular anachronism, but how many others might we have partially or fully missed? It has been comparably observed that certain elements of Shakespeare's message, as embodied in language peculiar to Elizabethan times, have been forever lost to posterity. Still, the possibility that moral truths have been obscured by time, or even by deliberate misrepresentation, should not be an insurmountable problem in discerning their essential relevance. Nor are the core verities of revealed religion any less true than those more mystically-derived. The fact that The Ten Commandments, for example, are virtually identical to the ethical precepts of the Buddha, should be a source of great encouragement. It highlights the common substrate of morality undergirding such divergent faiths disparately rooted in separate parts of the world.

The essential problem is that when we have veered off course in revealed religion, there is no ready means of recalibrating. I can say, for example, that the pepper shaker on the table is God, and, if in the words of Tertullian, "I believe because it is absurd," then neither I nor anyone else will ever be able to prove otherwise. It could be that the pepper shaker *is* God from the pantheistic viewpoint of a divine nature that suffuses all things. Yet even this epistemological twist would remain inaccessible, where faith excludes experience.

Where blind belief is the operative component of faith, the barrier between science and spirituality can only remain unassailable. And in a pragmatic world that largely abides by the Missouri adage of "show me," that approach will never fly, except among the credulous. If we hope to survive in a world where science is still precariously unguided by a comparably-elaborated ethics, such an obscene division and effective capitulation to ignorance, on matters of the most substantive human importance, cannot remain for long. The specific advantage of such experiential religions as Hinduism, Buddhism, and Taoism is that they embrace truths which, while in many ways equivalent to those of Christianity and Islam in their less doctrinaire aspects, are only *provisionally* accepted on faith prior to being substantiated, ultimately, by a heightened level of personal experience. In accordance with this approach, and as encapsulated by the Buddha in his dying words:

> Do not believe in anything simply because you have heard it. Do not believe in anything simply because it is spoken and rumored by many. Do not believe in anything simply because it is found written in your religious books. Do not believe in anything merely on the authority of your teachers and elders. Do not believe in traditions because they have been handed down for many generations. But after observation and analysis, when you find that anything agrees with reason and is conducive to the good and benefit of one and all, then accept it and live up to it (qtd in "Words of Wisdom).

Such a reasoned road to truth and to higher understanding, rather than the blind and irrational faith of Tertullian, is one which scientific minds can resonate with, and which Christianity in its contemporary decline would do well to adopt as its own, as religion, life, and the world become more integrative.

NOTES

1. Jesus prayed before his crucifixion, ". . . My Father, if it is not possible for this cup to be taken away unless I drink it, may your will be done" (Matthew 26.42).
2. "This is probably the best-known quotation by Karl Marx, the German economist and Communist political philosopher. The origin German text, in Critique of Hegel's Philosophy of Right, 1843 is: 'Die Religion . . . ist das Opium des Volkes.' This has been translated variously as 'religion is the opiate of the masses,' 'religion is the opium of the masses' and, in a version which German scholars prefer 'religion is the opium of the people.' The context the phrase appears is this: 'Religion is the sigh of the oppressed creature, the heart of a heartless world, just as it is the spirit of a spiritless situation. It is the opium of the people'" (Martin).

Chapter Ten

Projections and Conclusions

The onward march of technology threatens dire consequences for the human race if that technology is not soundly governed and its knowledge by wisdom completed. Still, there are several immediate steps we can take to prevent the worst from happening. We can develop a feeling for the larger humane context in which even our most competitive strivings—national, individual, and commercial—unfold. We can work to eliminate the hotbed of violence in conditions of basic need, and we can act to resolve the crucial dilemmas of unbridled power and of the inappropriate transfer of knowledge which perpetually threaten the peace and stability of the world. We will cover all of these areas and the problems that relate to them shortly.

Yet, should we accomplish all of these aims with a lapidary's precision, the result would still fall short of the good we could otherwise imagine. Utopia may never be possible, and, as perfection would never challenge mankind, neither may it be desirable. However, given the heightened degree of interdependence and of mutual influences which characterize the modern world, all that we may need to do is to establish an environment that is prevailingly peaceful rather than predominantly violent. As the world grows more interconnected, the unity we achieve should be one of peace rather than of destruction.

Should the aforementioned precautionary steps not be taken, the outcome for the near or distant future can only be one of two possibilities—neither of them desirable: 1) Whole portions of the earth, if not the entire planet, will be destroyed, if not by nuclear arms, then by some other hideous product of our wayward technological advance; 2) Democracy will be supplanted almost universally by dictatorship, as the need to maintain a level of control necessary to avert disaster becomes paramount. As oppressive forms of rule would only increase the tensions that ultimately lead to war, this makeshift solution would most likely only delay the inevitable. At the very least, dictatorship will perpetuate itself under conditions where its very establishment ensures that the root causes of violence, in societal malaise and collective discontent, will never be substantially addressed; thus to stave off a world filled with dead bodies, one would instead be created filled with "dead souls."

Such a dystopia could be expected to prevail, even here in the United States, with the passage of time and the advancement of technology, as the inevitable domestic concomitant of a conflicted international world "armed to the teeth." From the standpoint of our domestic development, the erosion of our core political values, as skewed in the direction of a gradually emerging despotism, has already been greatly abetted by several modern trends. These trends have exerted an unfortunate influence on the directions of our probable development, and have done so all at once.

The first and most consequential trend is the advancement of technology itself. It is worth noting, in this regard, that the Orwellian world of *1984* was not possible in 1948, the year in which, in a play on numbers, that prophetic novel was finalized. Nor was it pragmatically realizable in 1984 either. But it is chillingly possible today.

Ultra-modern surveillance technologies make it feasible to observe and record everything that occurs everywhere on the globe, simultaneously and in minute detail, in infrared or in Technicolor, while the power of cutting-edge computers provides for the ever-more-effective parsing and processing of the multitudinous inputs received. Such a real-time level of surveillance can now be accomplished at a speed and efficiency far beyond the capabilities of any current or former group of human beings, whatever their kind or malicious intentions. While this capability is being continuously modernized and refined to meet the ever-more-stringent demands of security in an ever-more-complex world, all it would take would be someone, somewhere along that progression, with the paranoia of a Richard Nixon, or worse, of a Josef Stalin, operating without sufficient constraints, to gather into one hand the scattered threads of this neatly prepackaged technology and to thereby turn us subtly, yet inexorably, in the direction of a darker governance.

Such technological enhancements have also been dangerously interwoven since the end of World War II with what President Eisenhower, in his departing address, described as "the military industrial complex."[1] This is the second consequential force implicated in our prospective—and to some degree current—decline. Here the will to profit and the will to power have been for the first time in our history dangerously and comprehensively conjoined. This has created a built-in and pernicious incentive for conducting limited wars, for arming for a potentially unlimited one, and for being impatient with what must appear, from a military or security standpoint, to be the deficiencies, weaknesses, and encumbrances of republican government.

Heading into World War II, the United States hardly had a military to speak of. As Omar Bradley described the anemic condition of the U.S. Army at the time of the fall of France:

> There were no organized basic training camps for draftees. Recruits were sent directly to existing or organizing Regular Army units for basic training, even though those units might be engaged in maneuvers. The Guard units—organized on paper into eighteen divisions—were ill-equipped and in some instances so ill-trained that the officers in charge had not the slightest idea of their jobs or how to train the men in their units. There was as yet no equipment for the newly recruited men. The much-publicized photos of recruits carrying broomsticks for rifles or using stovepipes to simulate artillery—and the slogan "Hurry up and wait"—were all to the point (91).

America emerged from World War II, in stark contrast to this earlier condition, strategically preeminent, technologically-armed, and with a pivotal role to play on the world stage. Its military-industrial complex, that which had made victory possible in the war against Germany and Japan, not only retained but exponentially expanded its role during the Cold War period that followed. It thus found its distinct and disharmonious inflection as an inordinately influential voice in U.S. policy.

In looking back even further, America's involvement in World War I spoke in clear and no uncertain terms of a waxing internationalism. As late as the 1930s, we were able to view ourselves as separate from the Old World with its characteristically more cynical regional outlook and discouragingly dimmed sense of idealism, from its narrowed horizons and its dark and skeptical retreat into the realm of hardened political realities. This regional dichotomy was no more apparent than from the circumstances surrounding the Versailles Treaty, in the literally and figuratively oceanic gap separating Woodrow Wilson's New World values and priorities from those of his skeptical and vindictive European partners.

The values that Wilson brought with him to Europe were the optimistic virtues of a true political republic, ones that had distinctively developed in the U.S. up to that time during what historians have more recently termed the period of "salutary neglect." During that protracted interval, America uniquely benefitted from its vast geographical separation from Europe. The European context contrastingly referred to was one characterized by the uneasy juxtaposition of erstwhile enemies, each more armed and powerful individually than the U.S. was at that time, each eyeing the others with fear over closely aligned borders. All continually faced far more complex balance of power issues than America, in its blissful isolation, ever had to contend with. Meanwhile, the American democratic trial, like a plant neatly hedged in, was allowed to develop independently, and in a state of almost perfect experimental isolation.

After World War I, Isolationism per se was still able to exert its influence as a more or less serious political force, one espoused, when the time came, most preeminently by flight hero Charles Lindbergh in his vocal opposition to America's entry into World War II. The issue of isolation versus entry was one ultimately decided, not by any domestic debate, but unceremoniously by the Japanese through their surprise attack on Pearl Harbor. By the end of World War II, Isolationism was a political view that could only be seriously entertained by the politically blind. Simultaneous with its demise as an effective political force, though perhaps not as well noted, came the end of isolation in the American democratic experiment. This is the next of our trends to be thoughtfully considered.

The democratic sapling that was planted here is no longer neatly hedged in. It is now but partially grown, and has, since the 1930s, been newly and fully exposed to the high winds of the world and its greater problems. The question that thus arises, though largely unarticulated as anathema to our political heritage, is whether a true republic can hope to survive at all under such dramatically changed conditions. As the environment we currently exist in is increasingly that of an integrative world, the answer is one that will depend on our grasp of that world's changed requirements.

By being holistic in our perspective, as part of this larger world, we can yet hope to retain the correspondingly advanced and integrative virtues of a genuine republic. By being conflicted and divided in relation to the reality beyond our borders, amidst ongoing and fearful attempts to continually defend ourselves from our now (through technology's leveraging role) uncomfortably close enemies, there can be no other end, ultimately, than to substantially abort our unique democratic experiment in the name of security. This capitulation would occur inevitably as part of a process that seems to already have started. It has threatened to unfold amidst: 1) the need to defend ourselves in a world where technology has given our enemies a far greater reach, where; 2) vast devastation on an increasingly broad scale no longer takes an army, but perhaps one individual alone, armed with a single weapon of mass destruction, and where; 3) so long as there is a substantial disparity in international living standards and in the means of satisfying collective aspirations, there will remain the will to terror and war. All of these forces and factors together make a good case for a creeping despotism, giving it motive, means, and opportunity combined. We have seen an inkling of this modern trend in both the McCarthyism of the 1950s, as a reflex of the Cold War, and in the more recent erosion of habeas corpus associated with the Patriot Act. Yet such responses are merely a hint of what is increasingly likely to occur if a now more unified world is plunged deeply and directly into violence.

The experience of the last great world war remains significant for mankind in two fundamental ways: 1) It gave us an extended preview of what total war in the current age of advanced technology would be like. The assembly line genocide of the concentration camps and the tragic nightmare of Hiroshima and Nagasaki linger hauntingly in our collective memory, and serve as pointed reminders of our enhanced capacity for destruction. 2) It marked the last instance in which total war between first-level powers, using all the weapons at their disposal, could be pursued in a way consistent with planetary survival. World War I, and to a progressively greater extent, World War II were, as a consequence of advancements in technology, unnaturally destructive wars. World War III would be so unnaturally destructive as to obliterate Nature and preclude the continued survival of Man. While the development of atomic weapons by the Soviets and of thermonuclear arms by both sides were, at the conclusion of the war, still some years away, the preliminary stage had already been set for a type of war that was no longer winnable.

At the height of the Cold War, international tensions built until, with the onset of the Cuban Missile Crisis, a critical flashpoint was reached. Yet instead of going to war, as any other adversaries at any other previous time would have done, both sides, peering into the abyss, turned away from it; a hotline agreement was signed shortly thereafter. It was followed by more significant and progressively more comprehensive agreements to limit the testing and deployment of nuclear arms (Morris 496-7).

In the aftermath of the Cuban Missile Crisis, a pattern of relations developed between the U.S. and the Soviet Union in which direct conflict was systematically avoided. The populations of both sides grew gradually habituated to a climate of peace and to living with "the bomb." The perpetuation of this climate facilitated the shift from military confrontation to economic competition.

As the threat from nuclear and other weapons of mass destruction shifts from superpower confrontation to second and third world terrorism, the angst and instability of developing nations comes more into focus as a category of world problem demanding an international solution. This has occurred amidst the correspondingly pressing need for the end of the inchworm to move steadily forward.

Technology, in both its positive and negative manifestations, has put such nations as Iran and Iraq, Saudi Arabia and Afghanistan, effectively in our backyard and us in theirs. This has made their destinies, however distinct their viewpoints or inherited culture may be from ours, now clearly, abruptly, and uncomfortably intertwined with our own security and progress. The misery gap between rich and poor is becoming ever more pronounced, and we are finally compelled to consider the squalor of "the least of our brothers," if only to beautify the neighborhood we now intimately share with them.

At this point, it would be appropriate to ask, in echoing the question once posed by Lenin (and even earlier by Chernyshevsky), "What is to be done?"

Beyond a hot war's mass devastation and a Cold War's fear and terror, like Marley's ghost, this question has returned in prodding and agonized tones of warning once again. In light of the integrative progress already achieved, we may hope that the solution or solutions we currently arrive at will be more positive, humane and enduring than Lenin's.

One Easter egg in the hunt for truth can be found in the writings of Rousseau. As he states from his *Discourse on the Origin of Inequality*:

> Savage man and civilized man differ so much at the bottom of their hearts and in their inclinations, that what constitutes the supreme happiness of the one would reduce the other to despair. The first sighs for nothing but repose and liberty; he desires only to live, and to be exempt from labor; nay, the ataraxy of the most confirmed Stoic falls short of his profound indifference to every other object. Civilized man, on the other hand, is always in motion, perpetually sweating and toiling, and racking his brains to find out occupations still more laborious: he continues a drudge to his last minute; nay, he courts death to be able to live, or renounces life to acquire immortality. He pays court to men in power whom he hates, and to rich men whom he despises; he sticks at nothing to have the honor of serving them; he boasts proudly of his baseness and their protection; and proud of his slavery, he speaks with disdain of those who have not the honor of sharing it. What a spectacle must the painful and envied labors of a European minister of state form in the eyes of a Caribbean! How many cruel deaths would not this indolent savage prefer to such a horrid life, which very often is not even sweetened by the pleasure of doing good? But to see the purpose of so many cares, his mind would first have to affix some meaning to these words power and reputation; he should be apprized that there are men who set value on the way they are looked on by the rest of mankind, who know how to be happy and satisfied with themselves on the testimony of others rather than upon their own. In fact, the real source of all those differences is that the savage lives within himself, whereas social man, constantly outside himself, knows only how to live in the opinion of others; and it is, if I may say so, merely from their judgment of him that he derives the consciousness of his own existence. It is foreign to my subject to show how this disposition engenders so much indifference toward good and evil, notwithstanding such fine discourses on morality; how everything, being reduced to appearances, becomes mere art and mummery; honor, friendship, virtue, and often vice itself, of which we at last learn the secret of boasting; how, in short, ever asking others what we are, and never daring to ask ourselves, in the midst of so much philosophy, humanity and politeness, and such sublime moral codes, we have nothing but a deceitful and frivolous exterior, honor without virtue, reason without wisdom, and pleasure without happiness. It is sufficient that I have proved that this is certainly not the original state of man, and that it is merely the spirit of society, and the inequality which society engenders, that thus change and transform all our natural inclinations (244-5).

This elongated excerpt is not included to deny, affirm, or critique Rousseau's political philosophy, but to benefit from his unique insight in observations directly relevant to the travails of the modern world. Least insightful, by far, is his depiction of savage (or natural) man. In fact, the natural state he describes, far from being a universal condition, has rarely existed at all outside of the fertile imaginations of certain Enlightenment thinkers, such as Rousseau himself, and in such remote regions as Mauritius, where even the hapless dodo bird managed to survive unmolested for centuries until with "the ataraxy of the most confirmed Stoic," it was slaughtered and rendered extinct. Rousseau's "noble savage," a sort of mellow antediluvian weed-smoker basking in stoic indolence, would, by the same measure, have been unceremoniously threshed in the Amazon rainforest or on the plains of the Serengeti. No, the strength of Rousseau's argument is definitely not there; it is in his revealing portrait of civilized man and the nature of the modern predicament.

As we traverse the nation's roads, we have conditioned ourselves to see "stop" as red and green as "go." This shared understanding is very useful overall; to ignore it, even occasionally, would be to court both legal retribution and death. Running a red light, unless we are drunk, insensible, reacting to a still more palpable threat (e.g. being pursued by an armed gang), or are otherwise temporarily or permanently insane, will spark, within us, a wholesome and legitimate fear. Yet, apart from the shared meaning that we have collectively—and arbitrarily—assigned to such alternating colors of the rainbow, there is nothing of "stop" in red, any more than there is of "go" in green. In fact, had we opted to tag them differently, such color codings could as easily have been reversed, much like the traffic flow along British and American roadways themselves.

Equally worth noting is that there is nothing inherently unnatural in attributing such values socially, as doing so actively promotes our survival; if everyone crossed the intersection at his or her own whim, our roadways would be shocking scenes of repetitive carnage. Nor, by the same token, is there anything inherently unnatural or corrupting about the formation of musical notes, circuit boards, or cities. Bringing such creations to life is simply what we do; they are as much of an extension of who we are as the building of nests, hives, or dams is for birds, bees, and beavers. And, although Rousseau himself directly contrasts the savage and civilized states, there is nothing inherently unnatural about civilization itself, nothing noble in the absence of flush toilets or artificial lighting, nor anything short of wondrous and admirable about socially-induced creativity and progress. It is only when civilization inherently contradicts Nature, a Nature whose underlying force and preponderant influence persist beneath the social veneer that hideous distortions and dire perils to our collective survival emerge.

In recent decades, we have fretted over the immersion of our youth into realities both virtual and violent. The remoteness of cyberspace from the realm of our more ordinary experience accentuates its isolation from physical

reality and that native reality's attendant social norms. The facility and eager-
ness with which, "like ducks to water" we plunge into such alternative realms
of fancy with a heart-pounding and thoroughgoing abandon has drawn attention to the cyber domain itself as a potential hotbed of incipient violence and
prime catalyst of immanent social decay. Yet the simple fact is, we have been
creating realities both virtual and violent all along, attaching such commonly
accepted terms as "nation," "society," and "property" to them. We have been
doing so all along without noticing; Rousseau was one of the few who
noticed.

The potential problem with any virtual reality, whether cyber-based or
social, is not that we have created it to begin with, or even necessarily that we
give ourselves to it with such heedless enthusiasm, but that we lose sight of
the underlying and more substantive reality of Life in the process. As both
our existence and our happiness continue to be naturally defined, the divergence between what is real and what is collectively imagined can lead to a
sense of alienation from life, and to a slew of social problems; such problems
typically arise at those contact points where, in Rousseau's own language, "it
is only from the opinions of others that civilized man derives the consciousness of his own existence."

Many prepackaged opinions are foisted on us as a concrete reality, a
reality alternative to and at times in defiance of Nature's. And although
Rousseau himself refrains from pursuing the matter directly, he acknowledges the indifference to truth, "to good and evil," that accrues from such a
temperament. In the civilized state, the gravitational center of meaning characteristically shifts from what is real to what is popular. The fact that we
jointly participate in and mutually reinforce our shared delusions infuses
them with an artificial veracity that, when filtered through a complex network of societal rewards and punishments, serves, in effect, to mimic and, in
the extreme, to substitute for a more naturally-based reality.

From the off-based and, to our eyes, comically artificial civilization of
Rousseau, where socialites in silk leggings and powdered wigs sneer at each
other across banquet halls, oblivious to any natural bond of friendship, to
"the pleasure of doing good," it is but one dark and degenerative step to the
dystopia of Orwell's 1984 (or to its real-world equivalent in Soviet "Socialist
Realism"[2]); here the natural reality of existence is not just partially but totally eclipsed by what society or the state would have us believe. What becomes
at issue is now no longer just an innocuous array of street lights. From the
shared meaning of signs, to the scrawl on a page that transforms itself into
language, to styles of dress and codes of conduct, to what we have, in the
broadest sense, elected to see as real, we have moved through ever-thicken-
ing clouds of individually benign customs, laws, and gestures, each successively blanketing the last until what is beneath is all but totally obscured,
engulfed by our own powerful imaginings; we are now in a position to not

only ignore, but to actually contradict truth in truth's own name, to violate the tenets of Nature, ours and those more general, without, like Condon's Manchurian candidate of Cold War fiction, so much as batting an eyelash in tacit recognition of the fact that this is what we have done. In Zen-like fashion, we have seamlessly become The Lie. We are enabled, by the virtual political reality that we have distractedly built, to love Big Brother[3] sincerely, whoever Big Brother may happen to be, with passionate tears in our eyes, bringing a false morality to perfection, conceiving it ideally from that vantage point where we first cease to recognize what is real or who we are. Of course, we are also then, inevitably, and not at all surprisingly, profoundly confused and unhappy, glutted to the brim of our false, twisted, and bloated egos with so much "honor without virtue, reason without wisdom, and pleasure without happiness," with everything superficial, contrived, postured and false. Enter nuclear arms.

Where in Orwell's 1984, "War is Peace" and "Freedom is Slavery," in the Cold War, "Annihilation is Victory,"—or in the equivalent refrain of the times, "better dead than red"—became the verbal launch code to an immanent planetary ruin. We, at that time, both the Soviets and ourselves, gained a nearly fatal and unnatural foothold in unreality, "groupthink," and broadbased collective imaginings to where we came to think about the unthinkable with serious intent. "Better dead than red" became, during the Cuban Missile Crisis, an actual working proposition among those military advisors surrounding JFK who had convinced themselves and others that a military attack on Cuba—one that would almost certainly have led to nuclear war—was the only legitimate response to Soviet aggression. Here the virtual reality of the nation, its conception of its enemies, how we narrowly defined them, and us in relation to them, nearly led to the supreme and unnatural illogic of planetary suicide. Such an imagined sum, a $2 - 2 =$ (victory for) 1 equation, could only have been arrived at from socially-imagined variables. It was perhaps only the keen insight of President Kennedy, an insight that allowed him to think with rare versatility "outside of the box," that almost singlehandedly saved us all from almost certain annihilation in the mindless "pied piper" march of weighty collective assumptions. Taking a page from Barbara Tuchman's *The Guns of August*, Kennedy saw how framing the current crisis within the assumptions of the last would lead inexorably, not to the fulfillment of the national interest, but to global annihilation.

The predominant historical lesson hanging thickly in the air at the time was that of the Munich Conference; feed the shark and it will come back meaner and hungrier. While that lesson retained its value, there were other contributing factors having an equal or greater bearing, most particularly, the unusual destructive capacity of those weapons of mass destruction each side had aimed at the other. It is insight that led Kennedy to see beyond the political and military assumptions of his times, much as Rousseau saw beyond the unnatural attributes of his own era's "civilized" environment. The development and application of such a mature insight itself, as a faculty now atrophied, nascent, or mistrusted in most moderns, may well be the key to our generation's own future survival.

While creating, enjoying, and benefiting from our virtual realities, we need to retain the natural and immediate connection with Life that allows us to see them for what they are. We need to pass the thread of our civilized progress through the eye of Nature's needle, without skewing our aim, forging a civilization aligned with Life, and not one opposed to it; the unvarnished truth will always persist beneath the veneer of our collective imaginings, reemerging in times of crisis with hideous retributive force to lightly trump our deluded actions and assumptions, however widely-honored they may be. We can still "give to Caesar what belongs to Caesar" (Mark 12.17) and retain the sense of Life as it is; but to do so properly, we need to restore our integral connection to Life, to those physical—and arguably metaphysical—laws, for which our prevailing group opinions in no sense substitute or negate.

In their book, *The Vortex: Where the Law of Attraction Assembles All Cooperative Relationships*, authors Esther and Jerry Hicks describe how we can reconnect with what they identify as Source, with the underlying energetic reality of the universe. We may do this, as they explain, by the internal processes of "aligning" and "allowing." Neither of these concepts immediately suggests any strenuous effort on our part, only a basic and natural recalibration with the truth of who we genuinely are. All of the world's great religious traditions have pointed to the primacy of this underlying connection with the core of life in determining our total success and happiness, while warning of the grave dysfunctions and misery that inevitably result from losing it. From Rousseau's civilized man, whose consciousness of his own existence derives only from the opinions of others, we venture into realms quantum with the following equivalent observation from *The Vortex*:

> Without exception, the flawed premise or unstable footing that most people stand on is because they care more about what someone else is thinking about them than how they themselves are feeling. So, over time, by interacting with many others . . . they have lost sight of their own Guidance and have become further and further separated from who-they-are. And so, they feel worse and worse as time goes on, and so they come to incorrect conclusion after incorrect conclusion until they are completely lost (Hicks and Hicks 9).

Civilized existence remains a subset of Nature. It would seem that those who recover their natural alignment with Life's central core might risk positioning themselves in direct and precarious opposition to the more arbitrary and self-serving standards of a largely opinion-based society. Yet, apart from those instances where the society itself is manifestly ill-founded, intrusive, or corrupt, as was the case with Nazi Germany or Soviet Russia, a very remarkable thing more typically occurs and may even occur unnoticed within the context of the more oppressive societies themselves; people actually function better within a civilized context—not just in a pristine, untrammeled, and natural condition—when in alignment with the more substantive reality that exists at all times beyond society itself; this is less surprising than it may at first seem, when we realize that the current of Truth continues to flow in subterranean persistence beneath the surface wash of whatever opinions may prevail at any given time. Those who are faithful to the truth will see more clearly, act more courageously, and be more in harmony with others as a result. There will be more in them to admire, and this will only serve, in most instances, to heighten their social effectiveness; the moral sacrifice of a Socrates or Jesus, contrary to those who would have us see morality itself as an impractical burden, is more the exception to a greater and more generally positive rule. Living in alignment (or coming into realignment) with life's underlying Source, is productive of, not only spiritual growth, but of expert functioning and positive results at all practical levels; this, in a nutshell, is the *Bhagavad Gita*'s central message of union with God as at once derived from and conducive to perfection in one's work.

Intuitive thought has been obliquely referenced in modern psychological parlance through the distinction between "left brain" and "right brain" functioning. Yet it remains to be noted that the thousand-petalled lotus of the mystic does not unfold over the right brain only. The socio-phobic techno-geek who designs attack lasers in his basement, and the daft flower child who can neither calculate nor earn her cab fare, are both victims of an imbalanced, non-integrative development—though our society in its own current and un-balanced state, is more likely to throw its own choicest garlands at the former, while denying them to the latter. Likewise, those who equate human intuition with the primitive instinct of animals, though it may serve a roughly equivalent function, by so stating, demonstrate their ignorance of its high and remarkable nature. As Christmas Humphreys states with reference to his "man of Zen": "he must be profoundly and actively aware of the limitations of the intellect, and conscious of the faculty which supercedes it, the intuition, by which alone the opposites are transcended, by which alone each man will find the subtle Middle Way which is above yet between them all, and KNOW reality" (6).

The *Tao Te Ching*, at once one of the simplest and most profound collections of truth-based aphorisms the world has ever known, offers its own clear and traditional statement of alignment: All things work together: / I have watched them reverting, / And have seen how they flourish/ And return again, each to his roots. / This I say, is the stillness: / A retreat to one's roots;/ Or better yet, return to the will of God, / Which is, I say, to constancy. / The knowledge of constancy / I call Enlightenment and say/ That not to know it /

Is blindness that works evil [my emphasis] (Tzu 77). Paramahansa Yogananda speaks in similar terms of "maya-bewildered men, the stumbling eyeless
of the world" (122). And finally, from the Bible: "they have eyes but do not
see" (Psalms 135-16). Lack of insight is a blindness, a blindness of the worst
and most potentially destructive kind. Restating the gist of a later-occurring,
yet fully consistent passage within the same edition of the *Tao Te Ching*,
translator Blakney paraphrases one of its particular passages as stating,
"Your perspicacity grows by small increments, so they say, but only by
trusting the light of the Way can true intelligence come to you." (Tzu 123).
To function without a developed intuition is to actually be missing a critical
component of intelligence, rendering the general thrust of human action, as
undertaken without it, skewed, warped, and destructive. What intuition is, at
basis, is the faculty of the soul's awareness. It connects the individual directly with Life and to its larger directions and dimensions. While our physical
senses, in their unenlightened form, assess only the landscape of what is
relevant to the demands of our personal survival, intuition has, as its reference point, that greater unity from which our personal existence derives and
to which it ultimately returns.

So what happens when all—or enough of us—and most particularly our
leadership—maintains a window on the Way (Tao) while operating within
the cubicle of his individual activities? Here the *Tao Te Ching* is explicit:
"When the Way rules the world, / Coach horses fertilize the fields; / When
the Way does not rule, / War horses breed in the parks" (Tzu 117). Albert
Einstein once said, "The unleashed power of the atom has changed everything except our thinking. Thus, we are drifting toward catastrophe beyond
conception. We shall require a substantially new manner of thinking if mankind is to survive." Perhaps, in alignment with life's (in a way) simplest, yet
deepest and most fundamental Truth, we can begin to change our thinking
now.

Each of us exists both as an individual and in relation to the larger whole.
These are the yin and yang, the complementary opposites of our existence.
Where we acknowledge both our individuality and our unity with the rest of
Life, we perceive most accurately and relate most morally; our barometer
becomes finely-attuned with that of our living environment. All of our social
ills originate from a basic violation of one or another of these basic life-
giving premises. Oppressive societies most clearly violate the first; those that
do damage to the environment, the second; those that economically exploit or
murder their own citizens do both. Even the soldier's sacrifice on the battlefield is rendered more noble by the hero's simultaneous acknowledgement of
both his own intrinsic worth and of the worthiness of his cause; the former
contributes value, the latter meaning. The only potential thread of ignobility
in such an otherwise admirable offering lies in its contradiction, should it so
contradict, of the still greater unity of Life, of the broader focus of sacrifice

and service represented by our intimate connection with our total environment, of our nation with other nations, and with humanity to all other forms of Life. Each of us is both unique and integral to our environment; this is not some vague sentiment to be touted only on rare occasions when parading our virtues seems glamorously advantageous, as in State of the Union addresses or appeals for cash donations, but as a very practical and underlying Truth that we routinely flout only at our peril, and at the accumulated cost of increasingly labyrinthine crises, such as the one that had been posed—and may yet be posed again—by nuclear arms.

Individually, what we need to do is to "stop the chatter," the delusive self-talk that shields us against any broader and more expansive outlook. We feel the tug of our own mindless opposition whenever our egotistical limits are challenged by a contrary way or truth. We need to soften the cacophony, and learn to listen intently once again, not to our own limited preconceptions, imaginings, pride, or fear, but to the clarifying quiet that ensues when the demand for what is vulgar and extraneous is put, at least for the moment, aside. We may then, like newborn infants resting upon our sheets, touch life whole and naked once again. We will assuredly re-don our social garments, as we interact with one another once more, weaving our way back through the customary avenues and symbols of society; but we will do so then with a heightened wisdom and savvy, for we will then be positioned to avoid, as a matter of course, confusing our trappings with their substance, our rituals with their truths, our concepts with their realities, the artificial constructs of nationhood with the genuine lives of the people. We will then hopefully emerge as individuals of mature insight, like JFK and Rousseau, no longer to be counted among "the stumbling eyeless of the world." We will then be empowered to aid fully and significantly in steering our collective course away from its treacherous waters.

As representatives of nations, we need to lift our preoccupied heads and take a reminding look at the stars. We are now and have forever been parts of a greater whole, living members of larger Life. It is against the backdrop of that larger and more inclusive identity that the worth of our efforts, both as individuals and as individual polities, is ultimately and beneficially defined. No one exists alone; neither does any nation, species, planet, or galaxy. It is the rest of life, of other people and peoples that accentuates our identities as unique individuals, while completing us. We are literally defined (i.e. defined as in the setting of finite limits to our identity) by what ostensibly opposes us, like the printing on this page, whose meaning is interpretable solely by its contrasting backdrop. Without the rest of the world being different, who we are would not even be recognizable. We should be grateful for this, and learn to profit from it, rather than to proceed by reacting violently to any variation or divergence that might potentially serve to show us a better way.

NOTES

1. Eisenhower's famous speech on the military industrial complex was delivered on January 17, 1961.

2. "Socialist realism was declared the reigning method of Soviet literature at The First All-Union Congress of Soviet Writers in August, 1934. The movement was enunciated by Maksim Gorky as a continuation of the Russian realist tradition best represented by Lev Tolstoi, infused with the ideology and optimism of socialism. It had roots in some pre-revolutionary intellectual circles, and was at least as appropriate as any other artistic method to express the vision of a socialist society. Socialist realism was unique only in that it was the sole official method of the state" (Geldern).

3. Symbol of totalitarian government in Orwell's 1984.

II

Challenges along the Way

Chapter Eleven

Problems of Perception and the Transfer of Knowledge

The analogy of the moving inchworm is exemplified through one of the core dilemmas in the advance of civilization, the use of the highest capacity in service to the basest instincts. It is a problem which centers on the discrepancy between natural abilities, and technological means. While the former appear to be relatively fixed and inalienable, the latter are readily transferable. In the advance of civilization, there are no safeguards comparable to those in nature to prevent the immature use of a mature capability. Still, without the ability to readily transfer knowledge, civilization itself would never have emerged.

It is possible, given human adaptability, for "the left hand not to know what the right hand is doing," for our thoughts to be neatly compartmentalized. In the military, soldiers are deliberately trained to focus on their operational objectives. For them to be distracted by philosophical concerns, when decisive action is needed, could well jeopardize not only their missions but their lives. Such selective processing of events has, however, been unduly abetted in modern times by perceptual distortions which demonstrate at many points, and in an ominous and troubling way, how our modern military technologies often dovetail poorly with our given instincts.

At the Battle of Bunker Hill, William Prescott, ordered his revolutionary soldiers not to fire on the advancing British until they saw "the whites of their eyes." Facing, witnessing, and interacting with one's enemy was an integral part of the conflict environment. During Desert Storm, TV viewers witnessed an A-10 fighter crew cheer as one of its missiles struck a target on the ground. It was only later that the public and pilots learned that the target destroyed was in fact a friendly vehicle and its occupants the victims of friendly fire (Hammer 33). Here, the nature of the weapons technology employed created such a wide disconnect that the observable action itself gave no hint as to its full and genuine consequences.

The process by which mankind, through technology's mediation, is alienated from the environment in which the results of our actions occur, can be traced in the development of both land and naval warfare. In the era of the old sailing ships, a battle at sea would typically climax with the boarding of the opposing ship by the winning party, perhaps even with hand-to-hand combat. By the World War I Battle of Jutland, the level of interactive separation had increased dramatically, causing the sense of perceptual immediacy to be correspondingly dimmed (Stevens and Westcott 98). Here, dreadnoughts scouted for each other across the misty waters of the North Sea, with the resulting line-of-sight engagements conducted at greater distances than the cannon of the old sailing vessels would allow (Wilmott and Pimlott 84).

By World War II, the pivotal role of the battleship came to be supplanted by that of the aircraft carrier. At the Battle of Midway, squadrons of aircraft were launched from the decks of opposing carriers that never came within visible range of each other. In the modern era, submarines armed with ballistic missiles patrol the ocean depths, evading detection, much as they had previously done at the height of the Cold War. Each reputedly possesses the ability to cripple the political and industrial infrastructure of any combatant nation from virtually any distance.

In modern times, the perceptual haze has not only intensified, but extended to envelop the civilian population. Treated to captivating visual displays of missile attacks during Desert Storm and in the follow-up Iraq War, TV audiences had almost to be reminded that what they were witnessing were the cataclysmic effects of battle and not the latest video game. Through the modern medium of television, one dramatic enough to be convincing, yet selective enough to be partial, most of us have been exposed from youth to prepackaged fictionalized portrayals of war. Absorbed in the dramatic visual panorama of colorful flags, exotic landscapes, unique clothing, and strange vehicles while we are lounging in front of the set with a sandwich and a beer, war to us can seem little less than glorious. Although we would protest to know better, we are not as likely as we should be to be as palpably convinced of war's harsh and desperate reality.

Frances Fitzgerald describes this experience, in relation to the Vietnam War, in the following way:

> The war was absurd for the civilians of both countries—but absurd in different ways. To one people the war would appear each day, compressed between advertisements and confined to a small space in the living room; the explosion of bombs and the cries of the wounded would become the background accompaniment to dinner. For the other people the war would come one day out of a clear blue sky. In a few minutes it would be over: the bombs, released by an

invisible pilot with incomprehensible intentions, would leave only the debris and dead behind (5).

Perceptual impediments, beyond those associated with conventional armaments, relate specifically to nuclear weapons. The minute before a nuclear strike, whether as part of an all-out exchange or in fulfillment of a terrorist threat, the target "under siege" is at peace; the minute after, little or nothing remains. In the midst of such an attack, there would be no word of troops on the march, no frantic messages passed along the battle line that the assault has finally commenced. Without our environment giving us clues, clues that our senses, instincts, and given natures can apprehend, we may be threatened and not be aware of it until that threat is actualized. Nor are we as likely to take appropriate measures to avert disaster in the absence of a palpable menace; all we may hear beforehand is the chirping of the birds in the trees.

While separation from the perceptual environment was part and parcel of the bomber pilot's experience in World War II, Korea, Vietnam, and elsewhere, the nuclear bomb is perhaps the ultimate symbol of that disconnect. It has so emerged as the potential means by which our radical separation from our living environment is made complete. That several ounces of pressure, exerted on a button, or more accurately on a series of keys, can produce a five megaton explosion 5,000 miles away defies our natural perception of the world. The man at "the button" is abstracted from reality, divorced from the consequences of his actions through the mediation of the mechanized process. Pressing the nuclear button becomes an almost symbolic act; the mechanism then takes over, becoming an ineluctable force for ruin in and of itself. Once Man's initial role of "button-presser" is enacted, a series of cataclysmic events is initiated over which he has no further control.

The elimination of Man from the equation of modern warfare that tabulates to his extinction was the theme of the 1969 film *Colossus: The Forbin Project,* which depicted a computer assuming control over the combined defense systems of The United States and the Soviet Union, threatening the world with nuclear devastation. Such a threat made comprehensively real was the fictional premise of the more recent *Terminator* films. The self-operation of weapons systems, more feasible as technology advances, heralds the final elimination of Man from the mechanized process of destruction that darkly determines his destiny. It does this by means of divesting him of his crucial decision-making role. Thus, initially, we are deprived of our ability to perceive, and in the end of our ability to act in favor of our survival.

In the Dalai Lama's book, *The Art of Happiness*, coauthor, Howard Cutler, M.D., references the work of Dr. Paul Brand in noting the bizarre yet typical response on the part of leprosy victims to the spread of the dire disease throughout their extremities. In not feeling the pain in their arms and legs, they come to regard these sensations as occurring to someone else, or as happening somewhere else, other than within their own bodies. It is as if their own appendages had become, amidst the unwholesome ravaging of their bodies and concomitant dulling of their senses, something completely alien to them (207-10). In a similar way, our technologically-induced anesthetization to the impact and consequences of modern weaponry is very problematic indeed, in light of its increasingly dire potential consequences. This is a troubling syndrome that makes moving beyond war the classic Clausewitzian cast as an acceptable and routine policy option, and the further strengthening of the international ties that bind, all the more imperative.

Still, despite our demonstrated capacity to selectively process reality, it can be argued that the same penetrating insight which allows great theoreticians, such as Einstein, to acquire knowledge of the subatomic universe, would unavoidably instill in them a corresponding awareness of the moral laws which rightly govern its use. At that mystical level where the peaks of knowledge are scaled, the individual man, or so it would seem, would not perceive himself as separate from others or his strivings apart from their benefit. The deepest understanding would thus be reserved ideally, by Nature, with its basic safeguards intact, only for those naturally prepared to use it well.

The pursuit of understanding for its own sake, or for the greater good of humanity, as the soundest motivation for scientific research, has also been identified, as exemplified by the philosophy of Plotinus, as an avenue to spiritual enlightenment. The Hindu *Upanishads* likewise contrast the road of knowledge directly with that of pleasure, acknowledging the fact that the individual, in choosing pleasure over knowledge opts to recognize the truth only when it suits him (Prabhavanda 16-17). The irony of choosing pleasure over knowledge, giving the lesser end the higher priority, is that the individual loses sight of how pleasure itself is to be knowledgeably obtained. When sick, the man of pleasure may try to evade his illness by postponing his trip to the doctor; he is thus more likely to turn an acute or reversible condition into a chronic or irreversible one. In being willfully ignorant of the sources of pain, he is proportionately more likely to be startled and overwhelmed by pain's actual occurrence.

In making the highest knowledge a servant to the lowest ends and by transferring the Promethean flame of weapons technology to radical elements in volatile regions, regions that are lagging behind developmentally, the most advanced nations in the nuclear age can similarly come to grief. Yet, from a human perspective, the highest, most potentially destructive knowledge is difficult to acquire by the natural process of understanding where the pursuit of pleasure so deludes the individual as to render him ignorant of the world around him. It is the ability to transfer knowledge to those ill-equipped to use it well, coupled with the exponentially more destructive impact of modern technology itself, which makes the rapid expansion of knowledge in the civilized state problematic.

While great men of science, such as Einstein, may seek knowledge for its own sake, they have transferred the products of their learning to other individuals for whom power and not knowledge is the primary end. Even where the attitude of the theoretical scientist is not thoroughly altruistic or pure, in not being motivated in the same way as the typical hard-core power-seeker, he is not as likely to appreciate, at a visceral level, the darkness of the ends to which his heightened knowledge is directed. While many of science's leading lights have sought scholarly recognition and fame, few have sought to deliberately destroy. While many have seen their knowledge put to evil uses, few have actively sought understanding for the purpose of promoting a manifest evil, or for the ends of power alone.

There exists plenty of evidence, nonetheless, which indicates how disruptive such a transfer of knowledge can be, with the 9-11 attacks serving as a modern example. Here it is both significant and symbolic that the instrument of terror on 9-11, the passenger jet airplane, was the same technology that, in its routine application, has served to shrink vast distances, transporting members of other cultures vastly different from ours directly to our doorstep. It was also, as notably, a borrowed technology. The 9-11 terrorists, armed only with box cutters and bile, lacked an indigenous weapons capability sufficient to achieve their aims, and sought, as a result, to hijack our airplanes specifically, rather than, so much, their passengers, in a violent attempt to turn our own technology against us.

From a larger perspective, it may be significant here, as well, that Osama Bin Laden, himself a Saudi national, did not establish his base of operations in Saudi Arabia, whose leadership—if not so uniformly its general population—has benefited from close and continuous ties to the West—but in Sudan and Afghanistan, both desperately poor nations, with the latter described in 2001 by Peter Bergen in *Holy War Inc.*, as being "so poor the World Bank no longer registers its economic indicators." Here the principle of the moving inchworm is implicated, though not as the sole determining ingredient in the terror formula. Note, in this connection, famed war correspondent William Shirer's description of Afghanistan from the 1930s, for it is one that could just as easily have been rendered accurately as late as the early 1990s, beyond which point, the Taliban began (somewhat ironically from the standpoint of integrative progress) to foster unity at the exorbitant cost of a draconian orthodoxy:

> I had seen one of the last remote lands in Central Asia on the verge of being pushed into the twentieth century. I had had a glimpse of it the way it must have been for more than two millennia, since Darius and then Alexander had come to Kabul. A tribal society, primitive, savage, living off its flocks and

barren fields, roving up and down the mountain valleys on camels, asses, or on foot, its people spending most of their time in mere survival . . . fighting off or attacking hostile tribes and government tax collectors, fearless of death in a way I envied, illiterate, uncivilized to a Westerner, but conscious of a long and continuous history, handed down by word of mouth from generation to generation (18).

While hunting down and executing such terrorist leaders as Osama Bin Laden may satisfy the public outcry for justice, it will no more diminish the prospect of terror, than creating more defensive weapons will eliminate the incidence of war. The root causes of terror, as of war, must be more substantially addressed by reducing the collective motivations for them, as dangerous and atavistic forces to be reckoned with in an increasingly complex modern world. They are forces, nonetheless, out of sync with that world's fundamentally changed requirements. Since we will never abandon the prospect of victory while our enemies still prevail, we can hope to transcend such alternating states as victory and defeat only by moving beyond war itself, and by rising to an appropriately integrative level in our political affairs, to a functional modality beyond traditional conflict and toward a more unified world.

The 9-11 terror scenario says something significant in itself about the ability to preserve peace outside of an integrative context, for those who are left on the fringes or who are otherwise infringed upon—whether actually or in their own perceptions—will remain disgruntled. And if that unhappiness is left to fester, it will metastasize into a collective will to violence. Any unity thus attained, in the absence of a more general satisfaction, will be at best artificial and temporary. This phenomenon has already been considered within the context of the Versailles Treaty and the rise of Hitler in Germany; but much the same can be said about the current motivations for terror in the ultra-modern world.

In *The Lord of the Rings*, the Dark Lord Sauron, initially the greatest threat to the peace of the fanciful "Middle Earth," remained, at the end of Tolkien's trilogy, as but a wisp of the evil he once represented. Returning from fiction to life, we will never eliminate the force of ill-will or the underlying motivations for terror entirely. A wisp of them will always remain. Some people or groups will forever have occasion to be unhappy. But in depriving terror of its substance, in its popular support within the regions which nurture and house it, we may similarly hope to render it ineffectual.

Such hope has been reflected in a very different category of response by which we take a sustained and active interest in such backward regions as Afghanistan. To do as we have done recently in our overall involvement there, to build more modern-style schools and health facilities, giving the indigenous population a healthy sampling of modernization and its benefits.

It is most likely that such benefits will, on average, though not necessarily at the outset and openly, be positively and enthusiastically received (provided we do not allow our glaring differences to openly clash, or familiarity to breed contempt). It can thus be expected that those who, by their terrorist definition, risk overturning the applecart of mutually beneficial development would lose much of their indigenous support.

ISIS, for its part, and by similar reasoning, may be seen to have spawned from the anomie and dislocations of a war-disfigured Iraq. Its emergence reiterates the historical point once emphasized by Stalinism in relation to an earlier Tsarism: that the societal crop harvested can only be as favorable as the developmental soil that nurtures it will allow. Alternatively expressed, this is none other than "the principle of the moving inchworm" as an internal dynamic. For Russia and Iraq, there was no "salutary neglect," only warfare and invasion, a circumstance reflected in an ensuing climate, not of democratic norms, but of brutality.

The situation is, of course, more complex than that, and too complex, in fact, to allow for an immediate optimism; there are factors of regional clash and of religious rivalry that motivate apart from immediate conditions. These are forces and factors that have been subtly intertwined with each region's distinct cultural development over time. Each represents a largely self-sustaining component of the popular will. Still, the impact is likely to be great and positive of introducing pragmatic benefits whose utility can hardly be denied. Such an approach, in the modern era, where weapons of mass destruction prevail, is clearly the way to go. Islam and the West may make strange bedfellows for a while, and there are always those who will resist the need for change or for timely accommodation. But the rules of tolerance associated with the integrative phase should eventually allow us to, not only put up with each other, but to ultimately celebrate our differences and to prosper together.

While poverty remains an essential interlocking piece of the larger terrorist puzzle, it is hardly the whole of it. Yet what remains is not irrelevant to our current discussion either. Many of Al Qaeda's leaders, like bin Laden himself, have been nowhere near to being uneducated or impoverished. The same made be said of ISIS and its leadership. What they clearly are—without exception—is violently resistant to the contemporary process of global integration itself. They view that process as inherently corrupting and threatening to their culture's core values. Their extreme reaction to it may be regarded as a birth pang in the emergence of the ultramodern world.

At the time of Jefferson's raid on the Barbary Pirates,[1] America was still distant influentially, and, by then prevailing means of travel, geographically from the Europe of our colonial progenitors, let alone from the more exotic inhabitants of the Near and Middle East. Through the long shadow we now cast, vastly extend through technology's enabling influence, our international presence alone, even as an ostensibly positive source of aid, was enough, at the time of the First Gulf War, to earn bin Laden's baleful notice, and to ignite Al Qaeda's enduring wrath.

True religion is not world-denying, nor, in a larger integrative sense, are physical and other worldly benefits, as exemplified by yogic science, isolated from the spiritual. The suicide bomber, who sacrifices his life in a wanton act of destruction, violates this integral relationship both from the standpoint of what he does to himself and by what he does to his victims. There is indeed something both keenly odd and poignantly tragic about any 9-11 scenario in which men in the prime of their lives are willing to crash airplanes into buildings to satisfy old men's political agendas. Such lives themselves must not be viewed as having much to offer to be squandered so readily. Even if the sacrifice is made to somehow benefit the remaining family or community, such groups, in vision or in truth, must be seen to be palpably deprived in some essential way, or to be missing a central element of basic humanity or morality, to make such an extreme and inhumane sacrifice seem like an acceptable option. What is clear here, if nothing else, about the whole terror scenario, whether as viewed from the standpoint of the suicidal perpetrators or from that of their innocent victims is that there are participating human beings who are relegated to a state of insignificance, people whose lives are violently excluded from the goals that the terrorist leaders envision. Such is the old way of divisiveness, exclusivity, brutality, and conflict we are collectively moving beyond.

As a faith that flourished in the desert, some of the harsher aspects of Islam can perhaps themselves be traced back to an experience in and with Nature, in which the clearest hope or advantage at one time seemed to lie beyond the horizon of immediate experience (or of this terrestrial life entirely). Such a radical view, or the continuing basis for it, then became permanently woven into the cultural fabric, despite any subsequent changes in the outward environment, including globalization. Still, if we allow ourselves to remain flexible, we can, as reflective human beings, retrace our steps, modernize our views, and make any valid religion what it is truly meant to be, a boon to life and an aid to human progress. As noted by Bernard Lewis:

> Islam is one of the world's great religions. It has given dignity and meaning to drab and impoverished lives. It has taught men of different races to live in brotherhood and people of different creeds to live side by side in reasonable tolerance. It inspired a great civilization in which others besides Muslims lived creative and useful lives and which, by its achievement, enriched the whole world. But Islam, like other religions, has also known periods when it inspired in some of its followers a mood of hatred and violence. It is our misfortune that we have to confront part of the Muslim world while it is going through such a period, and when most—though by no means all—of that hatred is directed against us (25).

So long as we know only one way of doing, only one way of being, there is a tendency to feel that our individual customs and approaches are absolute, and that they should be naturally followed and accepted by all. It is only through contrast that we are able to effectively compare both the positive and negative elements of our traditional understanding. As we develop toward an integrative world, technology, by bringing varied approaches within visible range, as it were, of each other encourages the development, corresponding to and promoted by its own advance, of a broad collective outlook. While young people may still select traditional values over more notably contemporary ones, they are now more likely to do so as a matter of choice, rather than in routine adherence to the way things have always been.

In a world that is now both richer in diversity and far more closely knit, citizens are increasingly capable of achieving wisdom in the Socratic sense of enhancing their understanding through an awareness of its limits. With technical knowledge expanding the vistas of what we have yet to understand in both the scientific and cultural spheres, we are less liable to feel that we know all that there is to know, or to see our way as the only way. Moreover, we are for the first time, both through our technologically-enhanced capacity for objective self-awareness, and through our comparably heightened ability to select from the best within competing systems of value, capable of a truly integrative understanding. Technology, in its relentless advance, thus contributes to the type of broad and humane vision which the development of technology itself necessitates, and which the modern world so desperately needs.

Knowledge, like the re-membered limbs of the Egyptian god Osiris,[2] may now be synthesized from the best of competing cultures, producing a quantum leap in our general understanding. Technology's positive side provides for the development of a broad collective vision. Technology's negative side, with the intensity of a cattle prod, now more forcefully compels it, with disaster waiting in the wings should we choose to tarry with error too long.

Through the ability to transfer knowledge, civilization, in its relentless advance, has bypassed the routine safeguards of Nature. It is as if adults had created matches so that children could play with them. Moreover, in societies where the ends of pleasure and power are implicitly given the highest priority, knowledge itself, as the superior end, is routinely subordinated to them. The compartmentalization of knowledge, so characteristic of modern life, likewise facilitates a willful ignorance and a routine disregard for action's larger consequences. These issues will now be more fully discussed.

NOTES

1. "Within days of his March 1801 inauguration as the third President of the United States, Thomas Jefferson ordered a naval and military expedition to North Africa, without the author-

ization of Congress, to put down regimes involved in slavery and piracy. The war was the first in which the U.S. flag was carried and planted overseas; it saw the baptism by fire of the U.S. Marine Corps—whose anthem boasts of action on 'the shores of Tripoli'—and it prefigured later struggles with both terrorism and jihad" (Hitchens).

2. "Osiris, in ancient Egyptian mythology, is a leading deity associated with fertility and vegetation; he personifies the principle of spiritual rebirth. Legend has it that Osiris' body was dismembered by his evil brother Set and his members scattered throughout the kingdom. All of the scattered parts with the exception of the phallus were later recovered by Osiris' wife, the goddess Isis who embalmed Osiris and restored him to life" (Prury 202).

Chapter Twelve

The Problem of Selfish Individualism

One of the hallmarks of the integrative state is the fundamental unity of knowledge, for to know everything about one thing, one would have to know everything about everything, so related are the parts to the whole of life. Yet, amidst the exponential growth in knowledge and the corresponding glut of raw information, it is no longer feasible for any one individual, within the limits of his personal understanding, to encompass the entirety of learning. It is sufficiently challenging and generally more useful to be versed in one specialized discipline or at most in a limited few.

The fields of human endeavor, in becoming more specialized and comprehensive in their own separate spheres, have grown more dependent on each other. In fulfilling their specialized tasks with the least distraction, each has become like the termite queen, whose system as it evolved, became so totally geared to a single function—in the queen's case reproduction—that other functions, even basic functions, were transferred or excluded. In this particular analogy, derived from Nature, the queen lacks the most basic means of locomotion, and must be rolled from place to place by others within the colony (Maeterlink and Sutro 99-101).

If an individual wishes to devote all his time and energies to the specialized tasks of management, he must hire researchers and administrative assistants, transferring to them those parts of the originally unified task which he would otherwise have to perform himself. He must further rely on an invisible legion of farmers, butchers, grain suppliers, transport workers, shopkeepers, and delivery boys to provide him on call with his afternoon lunch. Without these participants playing their supporting roles, contributing the products of their knowledge and labor, he would have to interrupt his efforts at midday to till fields, hunt game, or plant corn. Then, there are the delivery services, stationary suppliers, and technical support specialists without which his business would not only cease to grow, but would come to a grinding halt. To view these functions as less essential to the accomplishment of the overall mission is to miss the critical point, that no one task can be successfully completed without the others.

While the advance of civilization is more than ever collectively based, in our essential rules of conduct we still tend to think and act in accordance with the outdated precepts of a savage individualism. The attitude of "I did this, I worked hard for it, and in enjoying the result I owe nothing to anyone else, in either taxes or consideration," is not only ungracious but untrue. While depending in a multitude of ways on each other, helped along at each critical phase of work and life, we often at the end, glutted and satisfied, turn to bite the many hands that fed us. We carve out vast tracks of earth for our personal use, then hold in contempt those who must do without them. While depending on the labor of others, we commonly view their tasks as inferior to our own and, what is more, regard it as a privilege, and a sacrifice on our part that our subordinates are permitted to serve us. We often do not appreciate—or even acknowledge—that, when we occupy a leadership role, others have little alternative but to dance in step with our tune, that they everyday quietly and unquestioningly obey the rules of a game which has not manifestly favored them. When unemployed or otherwise in distress, we often hold them in contempt for exhibiting a level of status which only we, the decision-makers, have the unrestricted ability to change.

As the advancement of civilization has brought with it the process of specialization, the ability to transfer knowledge has allowed us to divide each original task and then bring the end result together, complete, as it were, on the factory floor. This encourages the false impression that our role all along had been the only crucial one, and all others contingent. The modern transfer of knowledge has become so routine that we often take it for granted, not realizing that our ability to fulfill our roles is dependent on our interrelationships. We continue to view life in society, not as an essentially social function, but, in keeping with the savage etiquette of nature, as a thoroughly individual one. And in our neglect and abuse of others, we erode the basis for larger collective accomplishments.

Since ancient times, the growth of civilization has been dependent on an ability to cooperate and to transfer knowledge. As historian William McNeill notes of Mesopotamian civilization:

> The urban revolution depended not only on . . . technical improvements but also upon a social reorganization which permitted coordination of effort among large numbers of men. Without such coordination, specialization and the development of technical skills (which depended on specialization) could not go very far. Even more important, the irrigation, without which cultivation of the Tigris-Euphrates valley was impossible, could only be undertaken and maintained by larger-scale social action (6).

As the growth of civilization depends on the ability to transfer knowledge, with social existence made possible only through cooperation, an extreme and brutal selfishness is an intrinsic violation of those collective norms upon which civilized existence itself is based. We have reached a point in the modern world, with the dawn of the nuclear age, where the ultimate penalty for such a violation, for allowing our collective knowledge to serve the atavistic ends of anarchy and greed, is the precipitous decline of civilization to a level of savagery and scarcity at which such brutal and divisive behavior would most aptly apply. If the transfer of knowledge, in the civilized state, is, for the most part, unavoidable, then we each have a vested interest in ensuring that all are empowered to advance beyond the level of savagery. Thus, whoever holds the keys of knowledge will be more likely to guard them well.

The Buddhist notion that our individual actions all revolve within the context of a greater field of dependently generated activity helps us to maintain a proper holistic perspective in a modern world that tends to overly specialize. It reminds us of how dependent we are from birth on each other, on the proper balance of conditions in the environment that mutually sustain us, and on how the simplest function may be the very linchpin that keeps the universal fabric from unraveling. We are all one; we all count—and we should all know it.

Chapter Thirteen

The Problem of Accelerating Change

A great irony of our existence and persistent source of anxiety is our ongoing quest for security amidst changeable circumstances which, by nature, are incapable of providing it. While this condition may always have pertained, through the rapid changes associated with modern life, encapsulated in the term "Future Shock," coined by sociologist Alvin Toffler, we are now more sharply aware of it. While the anxiety we experience in grappling with change can be bothersome, the courses we take to avoid it and to mask our perception of it, can be truly disastrous. Despite the contrariness of our instincts, we must find a way to cope with change in a world where change is constant without being overwhelmed by it. Our awkward reactions to change, our impassioned denials of it, the visceral way we respond to change and to its proponents, as well as our delusively comforting retreat into the realm of easy solutions, can all jeopardize life and increase the incidence of violence in the world environment. A key element in the mitigation of conflict is thus the cultivation of a more realistic sense of security, and the willingness, where the capacity itself is not in doubt, to adapt flexibly and wisely to conditions of change.

In a world of kaleidoscopic transformations, we ourselves must change if only to remain the same in relative terms. Where dynamic forces have transformed our living environment, traditional means may not promote traditional ends, making a blind adherence to tradition unreasonable. A mindless or misplaced resistance to change, in defense of traditional values, may itself pose a threat to a society's security and future well-being. Here the Buddhist injunction to "develop a mind that clings to naught" (Goldstein 157-8) particularly applies, encapsulating a basic strategy for ensuring survival and steady progress amidst overwhelmingly protean conditions.

It is well worth noting, in this regard, that populations forced to operate under the chokehold of oppression, motivated by the need to regulate change, are not typically the most creative. Non-creative societies cannot remain on the cutting edge of world developments. This was particularly well-demonstrated, in modern times, by the failure of world communism. As with the brutality of Stalinist rule, repression in the name of security can precipitate, through slow rot, a society's ultimate demise. The end result of denying such change as is bound to occur inevitably is, not to prevent it, but to cause it to occur all at once, through a violent revolutionary burst, or through sharp and bloody coups, rather than through gradual evolutionary reforms. When the dam of resistance to change finally bursts, the resulting Jacobin fury may sweep away those very traditions which such resistance had aimed to preserve and which would otherwise have remained unthreatened.

A popular bumper sticker reads, "If guns were outlawed, only outlaws would own guns." Whether one agrees with this idiomatic view or not, a comparable argument can be made for permitting change to proceed naturally, for, as similarly stated, if all change were made revolutionary, only revolutionaries would make change. Here the extreme right and left of the political spectrum define and accentuate each other, polarizing society away from its integrative center.

It was argued earlier that violence in the international environment reduces nations to the Hobbesian level of anarchy and to the Darwinian level of conflict. The same dynamic applies to a society's internal condition. By making no accommodation to change, by brutally squelching its opponents, all moderate opposition, all advocates of a compromise solution are eliminated in the initial round. Carry the oppression further, and it will then routinely extirpate the more bungling, least organized of the regime's hardened opponents. What typically remains are the most virulent, treacherous, brutal, and uncompromising elements of the opposition. Like insects exposed to successive strains of insecticide, these elements will have been molded, through the successively brutal actions of the state, to be immune to any extraneous considerations of law, morality, or future attempts at compromise. Such were the Bolsheviks in Russia, and the Communist revolutionaries under Mao, along with other such organized movements subsequently patterned after them.

By denying free expression in opposing change, the state burns its bridges, severing its functional ties with its population. Its actions make illegal opposition the only opposition, and the only available source to which the disgruntled can turn for redress. By making its oppressive policies routine, the state institutionalizes the extreme polarization of society, guaranteeing that change, when it eventually comes, will be violently undertaken.

Even should a dystopian society emerge which, in the mode of Orwell's Oceania in *1984,* the oppressed are conditioned to love their oppressors, the result would not be one in which the creative potential of its people can be actualized or the society itself securely advance. And, so long as there remain within the competitive marketplace of international politics freer and more viable alternatives, that society, so intent on preserving its institutions intact, will be outpaced and compelled to change. Here the ultra-flexibility of the Eastern approach, one which sees change as not only unavoidable but continuous and inevitable, nips in the bud the type of clinging resistance which blocks acceptance of needed transformation, making its routine expression violent and irregular.

During the 1960s, the U.S. was swept by a wave of revolutionary ferment which expressed itself creatively through music and other cultural media. It also bore witness to the excesses of that change, to the burning of flags, to the abuse of drugs, and to a popular taste for anarchy. Yet, even after the tumult of the 1960s had subsided and the Right in the 1980s spearheaded a Thermidorian reaction to it, much of what the 1960s championed has seeped into our popular culture. Civil rights and women's rights are now central concerns of both major parties, minus their extremes in the Black Panthers and the burning of bras. Respect for the environment, no longer the exclusive province of environmental kooks and self-styled nature mystics, is now rife in everything from environmentally friendly diapers to a self-proclaimed Republican conservative campaigning as the environmental president. The bizarre preoccupations of a "me generation" subsisting on brewer's yeast, natural herbs, and desiccated liver, is now less radically expressed through the emergence of a more health-conscious population, and to a widespread interest in fitness, and in food with good nutritional value.

While America survived the 1960s, reemerging from the turmoil enhanced and renewed, Gorbachev's Soviet Union could not cope with the impact of far more basic changes initiated from within and above, for, as Lao Tzu notes in the *Tao Te Ching*, "the stiffest tree is the readiest for the axe" (Tzu 129). In contrast to the U.S., the Soviet Union had not developed the institutional mechanisms to cope with change on a routine basis. Where attempts at change had been invariably interpreted as a challenge to the system's authority, those changes which ultimately occurred predictably became so. The Soviet system, in short, developed in such a way as to render it inflexible, and thus incapable of the adaptations ultimately required of it.

Addressing the problem of change from the perspective of material life and its limitations, the Buddhist *Dhammapada* states, "The world does not know that we all must end here; but those who do know it, their quarrels cease at once" (129). In this concise statement, and in its blunt acknowledgement of change, lies concealed the prescription to end war. When we are properly and fully aware of our own physical impermanence, and of the ceaseless transformations that are relentlessly occurring from moment to moment in the visible world around us, we will naturally refrain from sacrificing life—ours or anyone else's—for the sake of that which is worth far less than life, and which, as we are then most likely to far more clearly recognize, we can never hope to permanently control or own.

Chapter Fourteen

Problems with the Commercial Culture

Much of the rhetoric of the business world has its counterpart in the violence of nature—"clawing up the corporate ladder," "swimming with the sharks without being eaten alive," "bleeding the competition white." Adam Smith's very system of free market competition has been repeatedly referred to as "natural law economics;" its analogy to basic processes of nature may help to explain its abundant success (qtd. in Faulkner 431). Yet, in the competitive marketplace, as in the realm of professional sports, rivalry does not proceed amidst total anarchy, but within a framework of mutually cooperative relationships—the civilized rules of the game.

When two friends sit down to play chess, each typically plays to the best of his ability; each is determined to win. Yet each continues to acknowledge his friendship with the other and to understand that the competition between them serves a higher cooperative end. By pitting his skills against the other, each is furnished with a goal, and, through its pursuit, a means of self-improvement.

In the Bible, Jesus is quoted as saying "I come not to bring peace, but a sword" (Matthew 10.34-36). One can hardly doubt that this was a pronouncement optimistically received by many of his followers as a collective call-to-arms against Rome's continuing domination. But its true meaning is not only far more subtle, it is one which penetrates to the very heart of conflict's true meaning and role under modern conditions.

It seems—and most likely is—part of a larger plan at times, that we "lock swords" with people who bring out the worst in our natures, who espouse what we inherently oppose, and who effectively push all our buttons. Such interpersonal conflict gives us the opportunity, should we skillfully respond, to see by the light that our "opponents" cast on our flaws, that we may be exorcised of them and be healed. It is at such critical and opportune junctures, and in such ideally confrontational situations, that we find ourselves standing at the crossroads, faced with the type of choice that defines who we are as ethical and moral beings.

We can choose to react unthinkingly with what the Buddhists describe as the "ignorant emotions" of blind attachment and aversion, attachment to our private claims to truth, and aversion to the new and the unknown. Or we can expand our natures in integrative fashion, accommodating—while not necessarily assuming—the opinions and priorities of those who ostensibly oppose us. Among the nations of the world, this equates to a fundamental choice between peace and its attendant prosperity, and war and its inevitable destructiveness.

Competition certainly has its role. But closer to the mark, and to the underlying principles of the new integrative environment, is an economy based on the principles of optimum service and customer-oriented performance. According to karmic law, wealth is based on generosity. In corporate terms, this does not, of course, imply handing over the farm to anyone who wants it, or acquiescing to the unwarranted ascendance of market rivals, but of simply providing good service. Companies that do not function by this rule accumulate an inevitable backlog of complaints which, by inhibiting their forward momentum, keep them preoccupied with non-productive tasks. On such a retrograde basis, they are characteristically self-doomed, much like the aforementioned totalitarian states which stifle their own progress and pave the road to their inevitable demise with the ways of oppression.

While the commercial culture is geared to the satisfaction of the consumer's needs, it also depends to a notable extent on his continued *dis-satisfaction*, for it operates on the problematic and essentially Malthusian assumption that one can never have enough. It services needs by fostering an attitude of deprivation in the very midst of abundance.

By the standards of the consumer culture, a company can never have enough market share, a salesman enough clients, or a millionaire enough dollars. In an environment of presumed scarcity, one can only hope to acquire these things by getting them from others similarly burdened by the perception that *they* don't have enough either; hence conflict arises and is perpetuated. The false scarcities of wealth encourage the real scarcities of poverty. In the stampede to satisfy superfluous demands, genuine needs are neglected. Inviting some and not others to the economic table, jeopardizes the integrative basis upon which cooperative social structures and collective prosperity rest.

In the final reckoning, all the goods purchased, all the wealth acquired, all the status realized, all the limited victories won, are as so many potato chips whose taste leaves one hankering for more. It is no accident that the Buddha, after pondering for years the source of human misery, identified it as self-centered craving, for the very foundation of the happiness promised by craving is eroded by craving itself. This is not to fault the pursuit of happiness in those pleasures that life can provide, only the life itself lost in anxious scrambling, and the true nobility of purpose sacrificed daily on the altar of petty concerns. In making consumption the ultimate goal, Man himself is consumed.

Yet it is not necessary—or even desirable—for us to puritanically muzzle our pursuit of pleasure or to revamp the consumer culture, only that we properly preserve the balance on which a civilized culture depends. We must place its activities within the context in which its benefits can be properly enjoyed, while the frustrations, friction, and conflict currently associated with its characteristically overweighed activities are thoughtfully mitigated.

Any culture which uses virtue to service vice, the highest means to promote the lowest ends, is, to the extent that it does so, on a retrograde course, for the result represents an inversion of the natural course of mankind's evolutionary development. The consequences of such an inversion, in the modern world, as previously noted, are deadly. To the extent that we become a society whose aim is to service the consumer culture, rather than being serviced by it, we are headed downward. If we are to advance in the fullest sense, we must keep the highest in human nature always within our view, and actively promote an environment in which it is rewarded and not punished. We should operate from the principles of service, excellence, and that non-greed which acknowledges the underlying abundance of modern life, and through which more such abundance can readily manifest.

In the philosophical perspective of Buddhism, briefly noted above, the central problem of life is identified as "dukkha," often loosely and uninformatively translated as suffering. It is more accurately characterized, as in Huston Smith's reference to its original Pali meaning, as a condition in which the axle of life is offset in relation to its wheel. The result is an inordinate degree of friction and tension, in short, a very bumpy ride (111-12).

In keeping with this view, it can be argued that food, sex, drink, and the enjoyment of material possessions are all among the legitimate pleasures of life (Ecclesiastes 5.17-18). Each has its proper role and measure. Yet when made to substitute for what should be the individual's primary emphasis, the pursuit of perfection and purpose through higher pursuits directly or indirectly spiritual (with work itself as a potential spiritual practice), each can be extremely destructive.

In a similar sense, our collective attention should be focused, in the transition to an integrative world, on the accomplishment of higher aims and achievements, and most particularly, on the actualization of the full spiritual, mental, and physical potential of each individual. While an environment of scarcity makes the attainment of security and the quest for food the overriding concerns, the satisfaction of these more rudimentary needs forms the basis for higher achievements which now, in the absence of scarcity, become the natural focus. While the satisfaction of our physical needs allows us to tentatively exist, the satisfaction of our spiritual requirements allows us to fully live. With the attainment of security and abundance, the tide shifts from conflict to creativity, and the spiritual phase of our evolutionary journey begins in earnest.

Still, conditions of perceived as much as genuine scarcity can keep humanity from rising collectively to its proper stature. The perpetuation of atavistic standards amidst new integrative conditions can, likewise, prevent society from entering the state of its fully civilized development. Where scarcity, whether real or imagined, is implied, we are prompted to acquire more than we need to account for anticipated shortages. In recent times, in the realms of both defense and consumption, the hording of luxuries beyond the satisfaction of genuine needs, and the stockpiling of arms beyond the legitimate demands of defense have kept mankind preoccupied with backward pursuits, and dangerously immersed in an atavistic mentality, thus delaying our advancement beyond them. Like the axle attached to the periphery of the wheel, the preoccupation with primitive first phase concerns amidst modern, second phase transformational conditions can contribute, now more than ever in the past, to dissatisfaction, friction and conflict.

Any system which artificially forms its "integrative" basis by leaving someone or something behind threatens its own continuance and is ultimately doomed to punch its own clock. A spiritual economics founded on the perception of abundance, and on actively utilizing everyone's creative contributions, which, in the intrinsic approach of the *Bhagavad Gita*, sees "success as an outcome not a goal," is an integral part of our integrative future. It is a choice of Quantum abundance over Malthusian scarcity, of Christ and Darwin over Darwin alone, of human nature in perspective as a spiritual and material package, rather than a focus on the latter alone—any verbal declarations of allegiance to a substantially neglected faith to the contrary.

The savage game of one-upmanship is an old and outdated one. The domination of one individual or group by another, of colony by colonizer, of women by men, of labor by management, of race by race, of the individual by the state, and the brutal exercise of an arbitrary authority are incompatible with a civilized advance that is more than ever collectively based. It is a lingering vestige of a more primitive phase of development, one that we are collectively evolving beyond.

In the worldwide political arena, despotism is giving way to democracy among the most advanced nations, as with many not so advanced who are faithfully following their lead. Within the private sector, progressive firms are awakening to the benefits of having employees who are viewed as associates, where the accomplishment of an institution's aims is properly viewed as a collective endeavor, where everyone's role counts. This is not to say that, in the modern world, all employees will be equally paid, or be equal in their hiring or firing, but that all are to be regarded as equal in their combined dedication to the task at hand.

The age of the sweatshop taskmaster who arbitrarily presides over a petty workplace tyranny is a relic of the past, yet one which still occurs often enough to be troubling. There are enough people left at all controlling levels who, ruled by their instincts, become drunk with their authority. They attribute their success to their savagery, rather than to their knowledge and dedication, savagery in one-upping the competition, savagery in keeping their subordinates in line.

The game of power, of brutality, of bullying and dictatorial control, whether expressed by political or business leaders, is none other than the savage game of the jungle. So long as Man in society continues to play by its rules, he will carry the jungle with him, and along with it, the collective instinct and taste for violence. The knowledge which is the crest jewel of civilization, can make this instinctive savagery more destructive; absolute power can make it absolutely so.

The concomitants of civilization, being the same as the forces which sustain it, are freedom and a minimal level of shared prosperity (which should be a simple given amidst our joint compulsory adherence to the implied tenets of a social contract that, for richer or for poorer, for better or for worse, binds our individual destinies). Together they preclude the exercise of an arbitrary control and, in the modern world, of an inappropriately absolute power. Those of the savage state are anarchy, the wielding of arbitrary authority, the perception of limited resources violently acquired, and the brutal domination of the weak by the strong. With modern civilization the sum of its individual lives, the remnants of a misplaced savagery should be weeded out at all levels, if we wish to eliminate them at the highest level where, in the modern age, they present the greatest threat.

The Buddhist perception that "life is dukkha," as worth elaborating on in this context, is not the outlook from the beginning or from the end of the human journey, but only from the end of the beginning. We embark initially on life's grand adventure into a wondrous realm of dazzling perceptions and scintillating pleasures, infused with all the naïve and youthful enthusiasm we can muster. Yet we inevitably discover that for every joy in this world of duality (or relativity) there exists a corresponding sorrow. To the extent that we unchangingly rely on an ever-changing world, we are sure to be ultimately disappointed. For each pleasure, there is a corresponding pain, if only the pain of that pleasure's absence, felt with the same intensity of emotion by which the initial pleasure itself was charged. Every peg on life's spinning wheel that is pushed up is, with the turning of that wheel, hammered down—

or sheared off. With the ecstasy of human love comes its unpredictability and its torments, its concomitants of denial, frustration, fear and loss. From youth comes age; from life comes death. From the joy of friends comes the absence of friends and the death of all those whose unique love can never be individually replaced. Each such event, so experienced, independent of a greater truth or oneness, mocks our attempts at happiness, while pain itself, in its monotonous, mundane, and darkly repetitive visage, can only be seen as what it is; suffering (dukkha) alone, in the final analysis, seems real.

With each such experience we successively mourn, then move on. We continue to "live" as before, yet never quite as before; fresh disillusionments newly scar the heart while subtly and beneficially transforming us, tapping us on the shoulder to finally wake us up. This is the second transformative phase, the "dark night of the soul" of St. John of the Cross. Those who term themselves "realists" stop themselves there, self-satisfied and proud in their greater wisdom over those who have yet to know "the thorns that grow upon this rose of life" (Peers 118-9). Yet they too remain fundamentally ignorant in their unwillingness or inability to move on. They content themselves with a perception of existence as unremittingly harsh, unfair, and, in a world of complex arms and repetitive violence, doomed by all apparent reckoning to eventually self-destruct. Yet, there are a rare few, like the Buddha, who in continuing to investigate further, discover at the end of their quest a higher meaning and value at a transcendent level of perception and understanding. They are the ones who are ultimately able to look back on the vicissitudes of their individual lives as having beneficially served them, as having tempered and strengthened them by casting them back upon their own independent resources. They find, in the end, that Life, not only in the midst of its apparent harshness, but most particularly through and because of it, has conducted them to a place in the heart of reality where they can once again rejoice in the world's outward beauty, while being inwardly suffused with a spirit of progress and unassailable optimism. This is not the childish optimism they had known before, but optimism derived from having tasted of the tree of knowledge of good and evil, and made the circuitous route home. In their attainment lies the higher realism of hope for the world, a hope that we can move beyond, and thus put an end to, those wars that would destroy us. We can turn the page of communal life now before us to embark on still greater and more positive journeys, the likes of which we cannot at present even begin to properly imagine. From a realm of perpetual violence, and ostensibly random change, such rare individuals extract the enduring value of a subtly, yet intrinsically ordered world. It is a value which can only be realized when we understand fully what it means to be human in all of its layered dimensions. This leads us to a discussion of the larger context itself.

Chapter Fifteen

The Problem of the Larger Context

Science has exposed the inadequacy of superstitious beliefs to conjure away the ills of Man. Yet in failing to find God in a test tube, it has too quickly dismissed the good with the bad in its insistence on an unattainable precision. It has deprecated traditional morality without providing anything which can replace it. The sharpness of its material focus, like that of a microscopic lens, has put the larger spiritual context out of range. We can no longer tell we are in the forest, though we can examine every cell of bark on each tree.

In an effort to fill the void created by faith's departure, modern man has often attributed to material science, a completeness which is not there, and a saving potential which it cannot currently provide. While science cannot justifiably assert what it does not know, it should not deny what it cannot, by its own objective standards, disclaim.

We now, as before, are compelled to seek wholeness without certainty. From the vantage point that our heightened knowledge has given us, we must formulate a corresponding vision, however falteringly defined, of our larger purposes. For we cannot glimpse our future by glancing backward along the road to conflict, originally travelled. Nor can we measure our ultimate potential by the yardstick of what we already—with certainty—know. We must shape our approach to understanding and to life in general accordingly, mounting the jewel of knowledge within the open setting of the Greater Unknown. We are likewise prodded by contemporary imperatives to adopt a new way of knowing, itself beyond the strictly sensory and empirical, one that conjoins the ultramodern approaches of quantum mechanics and the traditional spiritual methodologies of the past. It is by such means that the gap can be bridged, as it now needs to be, between the spheres of faith and science.

In the modern world, religious faith has itself been influenced by a pseudo-scientific emphasis on precision, one which obscures the larger context of Truth. Here the elements of belief have become like the stipulations of a legal document. Recite the rights words and in the right order and suddenly—one is "saved," and the man who walks out the chapel door is not the same as the one who walked in, though his penchant for alcohol, online porn, or the routine harassment of his neighbor, continues unabated.

It can be argued that before the notion of salvation was born, good men followed the road that led to it. Before it was precisely defined, its principles were more broadly adhered to. Christ himself claimed that "not he who says Lord, Lord, but he who does the work of My Father in heaven" is truly worthy of salvation (Matthew 7.21). He thereby held the life of truth superior to any conceptual formulation of it.

Yet there are many who would insist that salvation depends on just such a measured formulation of belief. They deny the practical truth that there are Muslims, Hindus, Buddhists, and Jews who, by their mode of life, are better practicing Christians than many who claim the title. A broader, less hidebound approach to faith, one which, in revering the larger context, acts more in accordance with it, and, in valuing substance over verbal trappings, can help provide the world, in the nuclear age, with what it now so desperately needs, a sense of commonality transcending the bounds of custom, of compassion superior in its tolerance to the limits of particularistic belief. It alone can keep Hindus and Muslims, Muslims and Jews, Jews and Gentiles off each others throats and away from the nuclear trigger. This is a most important step in the movement toward an integrative world. That faiths whose common essence is love instead inspire hatred by the way their doctrines are divisively defended, suggests something wrong, not with faith, but with the particularistic approach to it; here the substance of faith can only be realized within the context of a more universal Truth.

The inhabitants of the modern world, in their quest for meaning, are confronted, on the one hand, and as previously noted, by a science which, in its skillful grasp of particulars, appears presumptuously to know it all, and by a faith which in applying the former's meticulous approach to itself, is more likely than not to declare itself a mystery. Here, our approach to understanding is as much a reflection of our material environment as the clothing we wear or the food we eat. The focus of attention, in the modern world, unlike that of other cultures in other times, is outwardly directed toward a material environment which contains more of interest and, in large part because we have helped to shape it that way, a less obvious potential for harm. Our willingness to begin investigating the myriad mysteries of this world, in contrast to an earlier preponderant reliance on religious faith, may itself have come about as a result of a more congenial mode of life. Man now, in the modern age, is more likely to throw rockets at the moon than to throw rocks at it. While the characteristic blindness of the ancients was to view natural events as the product of supernatural influences, that of the modern world, one which results from our current controlling relationship to the environment, is to measure all in terms of precise, yet limited, material standards.

The material struggles of nature do no more than challenge the ability of an individual or species to survive in a particular environment. While life is typically seen to evolve from the lower to the higher, many creatures, such as the shark, have successfully endured the trials of life far longer than Man while retaining their primitive form. If, in the aftermath of a nuclear war, the cockroach *does* survive while Man does not, a natural simplicity and dullness will have been exalted over an intelligence out of sync with its own ends. We, more clearly now than ever in the past, cannot find meaning in material struggles alone, and be thus out of keeping with our own highest nature—not if we hope collectively to survive.

An element of life cannot be fully understood apart from its larger context. Violence in nature, which would seem only cruelty, can appear otherwise when perceived from the vantage point of a species need to survive. Adversarial relationships, like that between the puma and the deer in the American Southwest, actually appear cooperative when viewed from this broader angle. Where puma populations declined at the hands—and guns— of jittery livestock owners, deer populations grew unchecked, and through its own overgrazing the deer died off in droves. The key difference between the puma and deer in their environment, and Man in the nuclear age, is that the natural predator's destructiveness has natural limits, while Mankind's destructive potential is virtually unrestrained. As similarly observed by Lipton and Bhaerman:

> While we humans are, indeed, part of the web of life, we are, fortunately, perched atop the food chain. We no longer have natural predators and so, as more than one cynical philosopher has observed, we prey on one another. There is a distinct difference between the violence of hunting a deer, which is a natural process in the established web of life, and hunting a deer hunter, which is a behavior that falls far outside of Nature's inherent morality. Our fundamental preoccupation with violence as a way of life is truly a misinterpretation of Nature (120).

The thermonuclear threat posed by adversarial nations in the Cold War had been compared to two enemies occupying a narrow room, each trying to kill each other, with a cache of dynamite as the only available weapon. That sole weapon, if ignited, would have destroyed them both. Dumb violence has always been a part of nature; it takes human sophistication to threaten to unnaturally extend the scope of a vast hate-filled destructiveness to the bounds, potentially, of the ultimate tragedy—the extinction of all life on earth.

The threat which military conflict now poses to our planetary survival cannot be adequately appreciated without acknowledging the distinction between natural and social conditions, and between those of the modern world and those of the past. Technological developments, such as the hydrogen bomb, when viewed within the context of the total expansion of knowledge and its accompanying social changes, could likewise alter our perspective on the meaning of that particular technology. An assumption of a spiritual dimension, which science at present can neither prove nor disprove (and in the absence of certainty tends to discount), could similarly alter our perspective on life by enlarging its total context. Physical survival, no longer fully an end in itself, would then have to be considered within the larger context of mankind's overall development, the survival of the species in relation to the evolution of the soul.

In the modern world, in particular, the greatest danger in ignoring the larger integrative context while focusing like a laser on its individual processes is that it leads us to see ultimate, rather than contingent value in those aims and activities characteristic of a more primitive phase of development. Thus, we are more apt to glorify conflict for its own sake, measure our success in relation to the failure of others, and view money, power, and pride, in the absence of higher aims, as the ultimate good. At the extreme of this approach, the worst atrocities of the modern era were committed by one man, Adolf Hitler, who regarded the perpetual struggle of life against itself as the supreme value. Possessing many of the most advanced technological tools of its age, Germany, under Hitler, utilized them for ends out of keeping with genuine human progress, with horrific consequences manifested on a mechanized scale previously unknown. This would be, however, but a foretaste of what must inevitably prevail if we continue in primitive mode toward the still greater technological advances now emerging. It is such a one-sided "modernistic" view which alone allows for the abuse of a vast—though still limited—knowledge, and for the use of nuclear arms. While technology itself is an integrative force, what is now of utmost concern is how we perceive ourselves in light of its advance.

Science, far from being unavoidably fated to obscure the larger context, has, in key ways, illuminated it. While it might be expected, for example, that the conquest of other worlds would diminish the relative value of this one, the opposite has in fact occurred. The view from space has highlighted earth's uniqueness as the only planet within perceptible range capable of supporting human life. While spy satellites have been routinely placed at the extremities of earth's orbit, there remains something almost sacrilegious about viewing the realm of space as a potential battleground. The expansive philosophical outlook which the view from space encourages makes us particularly conscious of the fragile unity of life here below. And with space giving us a sense of expanding beyond our limits, those limits would seem, appropriately, to include those of our ethical development as well.

It is as if Man, in venturing into space, has donned his best dinner suit to make the best impression on whoever might be waiting there to greet him. Yet, as we often treat our unfamiliar guests more courteously than we treat our relatives, the tools of science, in our everyday earthly dealings, have too often served the all-too-familiar ends of our hates, fears, and competitive strivings, rather than our communal virtues. And science, here on earth, has made us feel, rather than duly humble, almost omnipotent.

A perfect life form, the ultimate conceivable end of evolution, and hence of life itself, would not only be intelligent, but infinitely so, not merely powerful, but immeasurably so, and so thoroughly perfect as to transcend the limits of time. For it would not be perfect were it limited by temporal restrictions. Not limited by temporal restrictions, it would have to exist now as fully as in the future. Having no limit, it would be impossible to measure or to perceive by senses whose task it is to apprehend the world in de-finite bits. This is no less than a general description of God.

Our human language furnishes a clue to this notion of a perfection beyond time. The term "imperfective," as in the Russian language, is typically applied to verbs whose action is continuous in time, "perfective" to those whose action is complete. The effort to duplicate perfection in time cannot be achieved, for existence in time places inherent limits on perfectibility—with temporal existence itself being one of them. A perfection beyond limits would be beyond duality, beyond contradiction, hence beyond the ceaseless conflict associated with any conditions of natural struggle. It could conceivably be attained by humankind only through progressive attunement with that perfect source of immutable value.

Yet to view the progressive movement of life only in terms of contentious Darwinian values takes us in a different direction entirely, yielding a radically different outcome, particularly now in the nuclear age. With Nature, per traditional Darwinism, selecting for "beauty and aggression," a perfect life form, by this measure, would manifest not only as aggressive, but infinitely so, not only remarkable, like the black widow spider, in the lures of its perfidious beauty, but thoroughly so. Christian cosmology has already identified such a being. It is called the Devil.

In Hinduism, the Chandogya Upanishad describes how the devas (gods) and asuras (demons) approached the sage Prajapati, seeking to learn the truth of the Self, the truth that would lead to happiness. With each successive round of inquiry, Prajapati issues a partial and incomplete answer. He early identifies the physical dimension with the Self, as part and parcel of the more comprehensive domain of experience. According to the legend, the king of the asuras, Virochana, accepts this partial answer as total. As related in *The Upanishads*:

> Virochana, quite sure that the Self is the body, went back to the godless and
> began to teach them the body alone is to be saved, the body alone is to be
> adored. He taught them that whoever lives for indulging the senses will find
> joy in this world and the next (Easwaran 144-52).

Indra, representing the gods, not satisfied with this partial and inadequate explanation, returned for a series of subsequent and deeper inquiries through which these fractional truths were further amplified. He finally satisfied himself only with the complete truth, the truth that incorporated the spiritual dimension as well.

This legendary tale in the quest for meaning and happiness is pertinent to our discussion of survival in the nuclear age in several ways. First, it demonstrates how one missing ingredient in the formula of truth can totally alter its composition. This makes the quest for understanding, thus described, reminiscent, in a way, of the contemporary relationship between Newtonian and Quantum physics. The first delineated events masterfully at the atomic level, but being incomplete in its explanatory relevance, fell apart disconcertingly at the sub-atomic level. It also, in a more direct and worrisome parallel with traditional Darwinism, demonstrates how, for higher beings, evaluating the significance of material actions, without reference to their deeper spiritual context, will lead inevitably, not to the survival and prosperity invariably sought, but rather to profound suffering, and perhaps even, by modern means, to our own extinction. Along our quantum evolutionary course, there is still a strenuous striving. There are still in play many formidable obstacles, and a herculean overcoming of odds, but directed now toward the objectives of our spiritual future, rather than to the narrowly-limited survival ends of our primordial past only.

To approach life from the perspective of the larger integrative context is to evaluate our actions in light of their total consequences, our beliefs in terms of their fullest substantive value, our understanding in relation to its broadest limits. It is, in short, to measure the worth of our activities in terms of their broadest, most complete signification. Life, in the integrative phase, that of true civilization, is one in which Man, as a species, moves beyond the embryonic darkness of his primitive state, and in putting that behind, develops his higher potential within the context of an integrative consciousness, amidst adherence to those specifically spiritual laws that directly pertain to it. The development of civilization is thus equal in essence to the unfolding of the inner life. The great civilized achievements in art, architecture, science, and literature are but outward projections of a process of growth which is essentially inward, it being no coincidence that their beauty inspires us much as the ideals of true religion do. Yet despite civilization's notable accomplishments, the consequences of our routine neglect of the larger and more integrative context continue to be felt in a contemporary approach to understanding that remains precariously partial and myopic.

While we claim as a society to value individuals who show concern for the larger context, we honor them only in the way we would honor martyrs. Their knowledge is a collective good from which all in general benefit, yet which few in particular seem willing to champion. A modern American college student, for example, majoring in a field which emphasizes method, such as accounting or finance, would typically find, upon graduation, a well-paying job with any one of a number of prestigious firms. A student majoring in such disciplines as philosophy and history, fields which emphasize the concerns of the larger context, might be fortunate to find a job, upon graduation, sorting the other's checks. He is more likely to be judged, not in terms of his own well-thought-out values and priorities, but by the values and priorities of those who, from his perspective, appear hardly to have thought at all. For such an individual, the perception of life, as it is, becomes more agonizing through awareness of its contrasting ideal, and everyday his own undervalued status reminds him of it. What William Whyte wrote in 1957 is as true today as it was then:

> Relatively speaking, the liberal arts man remains in the cellar. About the only kind of job he is seriously considered for at the outset is sales work, and others regard this as a dog job offered people unqualified for anything else. If he still doesn't get the point, the salary differential should drive it home; with very few exceptions he is offered less money than his classmates who majored in business administration or engineering (Whyte 113).

If we are not to punish people for their virtues, and deprive society, and by extension the world, of the now critical benefit of having citizens of vision and compassion, we should give weight to the idea that there are larger contextual values beyond those of the marketplace. We should encourage people to feel that such values are worth cultivating, and that the consequence of adhering to them will not be financial or social martyrdom. We should give material support to our liberal arts faculties, as fully as to our business and science institutes, and to the profession of teaching as a whole. The "worth" of human beings should be measured less in crass monetary terms, and the attitude encouraged that there is a dignity in everyone which is more than worthy of our respect, regardless of wealth or social status.

We should also encourage the process by which good and talented people become actively involved in public life, and begin to view public service as what it essentially is—a public good—rather than as a pernicious drain on other more productive activities. A fact too often ignored is that many talented people refuse to go into public life, as the financial rewards are not as great, while the headaches can be more severe. Politicians, hounded by the press, have their dirty laundry aired in public, run the gauntlet of election-year abuse, and are typically subject to higher standards and greater scrutiny than their business-world counterparts. To the successful business leader, his anonymity is but another facet of his overall job security, a level of security which, taken in sum, is arguably far greater than the average career politician's.

That we routinely question the ethics of our elected officials, while seeking ever less regulation of business activities, speaks volumes about our collective attitudes and priorities, of our emphasis on limited particulars and on our neglect of the larger context. It reflects the prevailing perception that the man of business, pursuing profit and his own good is doing something useful, while the public servant, working for the more broadly-defined interests of his constituency, is little more than a balance-sheet eyesore whose value is not clear and who must continually justify the expense of keeping himself in office.

The larger context encompasses the ultimate "whys" and not just the immediate "hows." By referring appropriately to it, we make decisions in the present, so as to not leave the future with insoluble problems. The Cuban Missile Crisis, which created an immediate and critical choice between national prestige and global survival, was close to being just such an insoluble problem. It so remained until both sides agreed to back off, recognizing, in effect, that the ominous predicament they found themselves in was itself the greater issue. In a world of shifting fortunes and perpetual insecurity, will any amount of money ever be enough for us to give equal weight, as a modern society, to higher concerns? How much longer will we contribute to putting the world at risk by lowering our ethical sights while heightening our technological accuracy?

According to recent census numbers, there are over 6.5 billion people living in the world today. Life—or so it would seem—is cheap. But is there a larger context? Consider the following from a Buddhist source:

> Imagine, for instance, an ocean, vast as the three-thousandfold universe. In the depths of this ocean lives a blind turtle that rises to the surface only once every century. To attain a human birth is rarer than the chance occurrence of the turtle surfacing to find its head inside a yoke drifting at random on the water's surface. . . . Compared with the number of beings in the animal kingdom, humans are like stars seen during the day as compared with stars seen at night. And the same ratio may be applied between animals and pretas, and again between pretas and the denizens of the realms of hell.
>
> This precious human existence is thus most rare and extremely meaningful (Rinpoche 32).

As considered from such a wiser and higher perspective, the perspective of the larger context, perhaps even the most despicable individual, the least successful, the most misguided, the harshest, or the most deliberately cruel, is, by virtue of his intrinsic humanity, worth more than we can even hope to know.

Chapter Sixteen

The Issue of
Collective Goods and "Bads"

The peaceful or violent state of the world can be compared to the calm or agitated condition of a body of water. While the effects of any disturbance are greater at its source, they ripple far and wide to a point where we can refer to the larger body itself as being either calm or agitated. While the condition of the water is reflected in its individual waves, the individual wave either contributes to or serves to counteract that prevailing condition.

Similarly, within the vast ocean of humanity, the effect of one individual's actions on another impacts horizontally on still greater numbers of people, and vertically well into the future. Often the son who is beaten as a child, becomes the father who beats his son. Those who are raised in environments of hatred and fear, may grow up hating and fearing. From Caligula to Stalin, from Hitler to Hussein, it is difficult to find a ruthless dictator who was the product of a loving, nurturing environment. And in this high-tech and progressively integrative world, where the impact of our actions extends even further than before, how many more interconnected environments would be affected by the impact of one such environment on one such individual? Indeed, it would be more appropriate to ask how many would not.

The roots of contemporary violence run deep into the past. The results of the Versailles Conference, as previously noted, helped to bring Hitler to power. Hitler's invasion of Russia helped make the postwar Soviet Union even more militant and defensive, causing it to falsely perceive in America but a new version of the old fascist aggressor. The impact of the Cold War, by which America challenged Soviet militarism, produced McCarthyism. McCarthyism ruined the lives of many good and innocent people. Hence, to presume that the series of events, known as World War II, ended with the terms of surrender, is to lose sight of the ripple effect which such happenings, often viewed as limited in time and space, increasingly have on each other and on our collective future. Should we eventually be destroyed by the nuclear weapon we developed to defeat Nazi Germany, it would be Hitler's ultimate revenge.

History itself, as a discipline, is ill-equipped to trace the subtle, yet extensive impact which simpler, less noteworthy events may nonetheless have on those which more noticeably affect the tenor of our lives. Nor is it necessary to view individuals as mindless billiard balls in a game of historical determinism to admit the basic observable fact that Man, who determines his environment, is in the return stroke determined by it, much as a one who casts overboard an anchor tethered to him is consequently dragged down. Our world is shaped by the ongoing impact of our individual and collective actions. To acknowledge the existence of collective goods, in such a world, is at basis to acknowledge what we are, i.e., denizens of a single small planet whose fate hinges on the character of our joint and separate actions.

Many of the higher virtues of civilization are collective virtues. Many of its highest goals involve one form or another of collective attainment. Efforts to halt the spread of nuclear arms invariably require cooperative action at the highest international level. To solve the thorniest dilemmas of civilization, we must bridge the gap between our narrow concerns and now more universal needs where the two are more than ever conjoined. As the scope and impact of our actions extends ever further than before, we must show more routine concern for the common good and for the needs of the larger context; concern for the impact of industry on the environment, for a nation's approach to defense on the survival prospects of mankind, for the impact of "private" actions on the public weal. We must do so, if we are to advance technologically, without risking what we have already achieved.

In the movie, *The Thirteenth Warrior*, Antonio Banderas' character recites a variation of a Muslim prayer: "For all we should have thought, and have not thought; for all we should have said, and have not said; for all we should have done, and have not done, I pray Thee God for forgiveness." In the modern world of enhanced interconnectedness, to remain at peace and to secure a survivable future, we must concern ourselves not only with what we think, say, and do, but with what we fail to think, say, or do, within the domain of our expanded influence.

The injustices we never think to remedy, leaving others to suffer them alone, the love never expressed by a parent to a child whose desolation festers into violence against his peers, the healing work neglected in decimated war-zones our nation leaves behind like a fornicator bolting out of bed, the suffering millions conveniently ignored, left in too much pain and for far too long to ask "why?," until all vestige of hope and humanity are bled clean from their veins; all of these exact their ultimate price in violence, terror, and war. All attest to human bonds severed and collective interests denied that, while they solidly exist, are just as persistently disowned.

This is a world that values peace, but which never seems able to enjoy much peace for long. If that basic circumstance is to change, we must learn to feel more than an obligatory outrage against the injustices and evils committed around us, as if our idyllic, yet artificial isolation were being rudely interrupted by them; we must continuously and proactively, individually and collectively, do our part.

Chapter Seventeen

The Problem of Basic Need

People do not turn to violence typically out of preference, but more out of frustration and desperation, or because an environment of violence and scarcity, one which effectively determines a belligerent response, is all they have ever known.

Violent action invariably entails risk—the risk of punishment, the risk of loss. Risk, most particularly when it jeopardizes one's very survival, is something which most change-resistant, comfort-and-security-loving human beings would wish more than anything else to avoid. One can localize the initial source of most violence, and particularly of most collective violence, in conditions of basic need.

It is nonetheless true that, when a momentum of violence has been established, violence is routinely chosen over better alternatives. By such time, the people involved have typically been reduced through the harsh terms of their existence to the level of wild beasts, and against such beasts there regrettably is no reasoning. As with the former Entente in relation to Hitler, the non-aggressor is compelled, at this latter juncture, to respond to the aggressor in his own unfavorable terms, and by means of negative rather than positive reinforcement. The price paid then is invariably higher than an earlier and timelier course, based on positive incentives, would have been.

In World War II, Hitler was defeated by the Allies, but only at the cost of tens of millions of lives. In America's inner cities, crime has been to some degree deterred through uniformed patrols and lengthy prison terms. Yet this brand of negative reinforcement exacts its notable cost as well. Routine patrols have cost many officers their lives, while prison upkeep costs taxpayers revenues which could have been productively spent servicing other needs.

The argument that violence is rooted in need is not presented as a rationale for violence, or to imply that human beings, either individually or collectively, are not responsible for their behavior. The essential difference between Man and animal is that Man, while possessing instincts, is not thoroughly bound by them. Man's greater ability to choose implies a heightened degree of personal and social responsibility.

The fact remains that conditions of need drag mankind, in the civilized state, back to the stage described by Darwin and post-referenced to Malthus of a struggle for scarce resources. In this stage, whether it prevails in a tropical rain forest or on the streets of the concrete jungle, violent modes of response are the norm. In light of this understanding, it makes no sense, from anyone's self-interested perspective, to relegate people to the wilderness, and then condemn them for being wild, or to upbraid them for a lack of values, where a lack of resources or opportunity is persistently rife.

A large part of terrorism, domestic or foreign, is outrage twisted into hate and into a cold determination to destroy. As such, it is incompatible with any sound moral purpose. Yet were we, for our part, to visibly and routinely focus on such problems as poverty, hopelessness, and lack of essential participation in the benefits of the global economy, vengeful acts of terror would lose much of their perceived validity. Incentive for them would be thereby dramatically reduced. The aim of the terrorist, like that of a spoiled child, is to shock us into paying attention by engaging in dramatically destructive actions. Where such attention is already being paid, terror can only serve to diminish our involvement or to render its character less favorable.

In *The Masque of the Red Death,* Edgar Allen Poe portrayed a land ravaged by plague. Those individuals fortunate enough to avoid the infection gathered together, sequestering themselves from the rest of the community. Momentarily safe in their fortified shelter, they proceeded to hold a raucous masquerade ball. The festivities continued unabated, while, beyond its protective walls, the plague ravaged on. As Poe describes the mood of the revelers assembled by the Prince Prospero:

> The external world could take care of itself. In the meantime it was folly to grieve, or to think. The prince had provided all the appliances of pleasure. There were buffoons, there were improvisatori, there were ballet-dancers, there were musicians, there was Beauty, there was wine. All these and security within. Without was the "Red Death" (175).

The revelers at Prospero's party did all that they could to blunt their own perception of the outside world and its problems. Yet finally, and, as it would seem, inevitably, someone afflicted with the disease, the personification of the Red Death itself, entered into their midst. The private party was finally over.

If, in the final analysis, it can be concluded that the fortunate bear no responsibility to the unfortunate, the enfranchised to the disenfranchised, or the developed nation to the underdeveloped world, the response necessitated by events would be just the same. Though our actions may not contribute to the plagues of poverty or injustice, our inaction would still serve to perpetuate them. Though we may not have contributed to the cause, we must, on the basis of self-interest alone, if for no greater or more compassionate reason, contribute to the eventual solution; for in an interconnected world we cannot help but be affected by the larger result.

In grappling with the issue of needs, social scientist, Abraham Maslow, spoke in terms of a systematic hierarchy. He recognized that, as one human need was met, another more elevated priority could be expected to take its place. This would propel the individual, in his drive for satisfaction, to move expansively onward toward higher and more enriching objectives.

A basic problem with Maslow's rational hierarchy stems from a consideration of our frequently more irrational nature. In *The Yoga of the Bhagavad Gita*, Sri Krishna Prem describes the backdrop to the battle for the human soul that this epic work allegorizes. Here the Kauravas represent the dark or Asurik forces, while their opponents, the Pandavas, personify the forces of light, truth, and goodness. As described by Krishna Prem:

> The Divinely-born Arjuna, with his four brothers, was brought up with his cousins, the Kauravas, at the Court of the latter's father, Dhritarashtra, the king who, though legally disqualified by his blindness, had seized and held the throne. Not content with the seizing of the throne, the old king did not even hold the balance evenly between his sons and their cousins, the Pandavas, but constantly favored the former. Hostility soon developed between the two parties and, after a brief attempt to divide the realm between them, the Pandavas were defeated at dice by trickery and made to wander for twelve long years in exile, followed by a thirteenth year in which their very whereabouts had to remain unknown. At the conclusion of this period the well-meaning but weak king found it impossible to persuade his headstrong and evil-minded son, Duryodhana, to restore to the Pandavas their share of the kingdom and, in spite of fruitless attempts to bring about a reconciliation by Sanjaya, Dhritarashtra's charioteer, by Bhishma, his wise counselor, and even by the Lord Krishna Himself, war could not be averted, and the rival hosts faced each other on the field of Kurukshetra. It is at this point that the Gita commences (xix-xx).

The Gita's symbolism has an instructive bearing on the elemental forces of human nature underlying a tragically-conflicted and war-plagued world. Here Dhritarashtra, the old king, is portrayed as "blind," while his son, Duryodhana, is characterized as "headstrong and evil-minded."

Divested of such allegorical cloaking, mindless greed and the lust for power, as more bluntly identified, are not merely goals in a rational hierarchy, but, as the *Bhagavad Gita* portrays them, blinding and obsessive pursuits. Operating under their spell is less like being misinformed, than it is like being drunk. Here the influence of the drive itself is so miasmic as to preclude, in many cases and in a tie-back to Maslow, any rational progression up the hierarchy of needs to what may in fact be good, better, or best for the individual, as logically construed. An essential problem with Maslow's hierarchy of needs thus stems from the all-too-evident fact that human beings are not always rational, *and* that the consequences of their irrationality are rarely limited to the sphere of their individual activities alone. Alexander, Napoleon, and Hitler were rational enough to pursue their empires systematically, but not rational enough to end that pursuit before it had gone too far. At a point where other needs—à la Maslow—should have rationally taken precedence, these warlords continued on in much the same belligerent mode as before, as their lust for conquest became a blinding and obsessive aim.

When you are hungry, your "need" would be for a slice of bread, or shred of meat to sustain you for another day. Yet, in social terms, a second yacht, or membership to an exclusive club, may also be considered a "need," as the individual, in his rambling pursuit of happiness, attempts to "keep up with the Jones's." The tragic irony is that the wealthy individual's less rudimentary need could be expected to take legislative precedence, amidst the perennial demand to fill the campaign coffers, over the poor man's need for a simple slice of bread.

What might you give to the man (or woman) who has everything? Nothing less than the assurance that what's been accumulated will not be taken away. This encapsulates, in a nutshell, the modern political preoccupation with tax cuts as a pivotal campaign issue—not the cogent version of the argument that those who have more wealth will have more wealth to invest (i.e., wealth as a rational and expansive pursuit)—but (irrationally) as political snake oil for all that ails us, whatever the harm done, and no matter what collective good is consequently denied.

There are some, like the Kennedys, who, after comprehensively satisfying their personal or familial goals of security and wealth, move on to more expansive aims that broadly intersect with the greater good of society. Still, for others, the siren calls of power, conquest, or material gain prevent them from switching rails. The end result, not of pursuing such limited goals, but of pursuing them in a lopsided way, is to leave the tail end of society's inchworm ever-further behind in a way that fosters social imbalance, making marginalization, conflict and violence routinely inevitable, much as it does in the *Bhagavad Gita* itself.

Chapter Eighteen

The Problem of Malignant Nationalism

Men are driven to the transcendence of social, as to religious, ideals through the consciousness of their mortality, as to the wretchedness of their lives. They seek value in that which will endure beyond them as the truth which can give their present lives meaning. The more evident the need to escape the bounds of suffering and the inevitability of death, the more emphatic becomes devotion to the ideal in its transcendent aspect.

When Rousseau claimed that "civilized" man would sacrifice his life for the sake of immortality, it is to the immortality of the state that his statement would, for the twentieth century, have most aptly pertained (Ritter and Bondanella 56). The nation-state has notoriously served as the focal point for those aspirations and hopes which the individual cannot fulfill due to his limitations and mortality. In the case of extreme nationalism, separate individuals merge their identities into that of the political collective in the hope of reconstituting them at a higher level. The power of the state gives it a freedom and scope of action which the individual man lacks, to the point of transcending the bounds of personal morality. It has, at times and as such, become the concentrated focus of mankind's darkest yearnings. While the individual is typically restrained by morality, law, conscience and fear from killing, the state has been known to kill on his behalf with impunity, and out of a sense of justice based solely on its need.

In medieval times, organized religion was the repository of those communal virtues on which European civilization was based. When Nietzsche proclaimed that "God is dead," he heralded the compensatory rise of the nation-state as the false god of the modern world. In more recent times, communist regimes have opposed the practice of organized religion so vehemently due largely to the fact that the idealized workers' state attempted to fulfill the same level of needs as had once been addressed by religion.

The nation-state furnishes a means of enhancing human identity beyond its natural limits. It gives idealized expression to the satisfaction of human needs beyond the more adverse conditions of their actual fulfillment. It represents, in both its positive and negative variants, a collective "immortality" and "truth" in opposition to Nature's harsh denial of individual existence and values. Yet, in being an intangible ideal, set apart from the tangible reality of life, it has, at times and as such, become the basis for further separating the individual man from his natural identity and purposes.

Where the State, as in fascism, comes to be viewed as the fullest expression of human identity, human needs are not typically met by it, but subordinated to it. The ritual sacrifice of human lives on the altar of primitive gods is little different from the modern sacrifice of men in their individual lives to the "needs" of the total state. As idealized through Hitler's benighted view, the State is everything—the individual nothing. This is the precise inverse of progressive government in the integrative state, contrastingly expressed by the precepts of our own founding fathers. Those precepts referred the government's power back to the authority of the people, whose lives were, and remain, its sole and true reality. This peculiar alienation of Man from his natural identity and purposes is central to the dilemma of violence in the nuclear age, for it alone allows us to collectively commit atrocities which, individually, we would not dream of considering, while still permitting us to act out of a presumed sense of righteousness. It alone would allow for the use of nuclear arms.

In the Biblical story of Mary Magdalene, Christ challenged each member of the mob "who (was) without sin, to cast the first stone" (John 8.7). In doing so, he pointedly shifted the attention of each participant away from his false, yet comforting, group identification, back to a blunt awareness of his true human identity. It is that identity which participation in a larger mass can often obscure. Such false identification with the crowd, whose role and identity dilute individual responsibility, is what led Nazi death camp guards to exculpate themselves at the Nuremberg Trials—as elsewhere—under the blanket moral insurance policy that they were "simply following orders."

The alienation of the man from himself, from his own perceptions, and from the defining elements of his own humanity, were echoed in the sepulcher silence of those Allied bomber pilots of World War II who visited the cities they destroyed, witnessing for the first time the carnage they inflicted; reflected in the tear-lined faces of ordinary German citizens who paraded past the corpses of concentration camp victims; reflected, as well, in the life of the average United States citizen who, though he would hesitate to swat a fly, routinely casts his ballot for the proponents of an enhanced militarism. All attest to that most insidious and peculiarly modern form of alienation, the alienation of a man from himself.

Whether with that bomber pilot whose payload kills unseen victims, the population whose radio provides the sole (and misleading) connection to the reality of its nation's atrocities, or the citizen who votes in a mechanical booth for policies he cannot sense or feel, it is technology which makes possible the alienation of man from his environment and from himself. And it is technology, misused through an incompleteness of knowledge, which alone can sustain the vast collective identity of a totalitarian state.

While such alienation is not unique to the technological environment, technology has enhanced the process by which our vision is routinely obscured, made the unfolding of that process particularly easy, and the consequences of technology's own abuse especially deadly. Technology, as with the many positive and integrative aspects of the Internet, as with the penetration of unauthorized Western media into countries more rigidly controlled, serves, in its true nature, as an integrative force, provided, once again, that the knowledge it represents finds completeness through wisdom's elevated guidance.

Ignorance, however blissful it may seem on the surface, is never a safe choice. In the modern world, its consequences can be especially deadly. Through both our personal and collective delusions, we deliberately skew our own perception of reality, reading infinite value into finite pursuits as a means of avoiding discomfort, and an immortality apart from the human spirit into the imagined life of the state. The jingoistic nationalist, as evaluated in these terms, is one who hides behind the shield of his own limited conceptions, oblivious to the impact that reality may have on them. While a particular Russian and a particular American may, for example, have more in common than one American to another, or of one Russian to another, a false concept of nationhood can erect artificial barriers between them. Such collective delusions run deeper than individual ones, involving as they do the phenomenon of "the emperor's new clothes."

In society, a lie can seem more valid than the truth, the more people share a belief in it. Where, as was the case with Galileo, people are penalized for views which are, in fact, correct, truth, through the consequences which society imposes, can be made to appear false. The abuse of technology can enhance this "virtual reality" of the state, making such collective delusions more credible.

Because those beliefs which constitute the core of distinct national identities are imagined largely apart from the actual conditions of life, it is nearly impossible to establish any practical meeting ground amongst them. How, for example, would we be able to reconcile the notion of Muslim jihad with that of communist world revolution? If any connection were to be established between them, it too would have to be imagined apart from the reality of basic needs. Due to the practical irreconcilability of such distinct value systems, imagined apart from the actual conditions of life, the potential for pointless conflict in the modern world is greatly enhanced.

In the United States, an advanced form of political order emerged designed to be "of the people, by the people, and for the people." In proceeding on this basis, it maintained a functional link with the reality of human existence and gave priority, at least in theory, to the satisfaction of human needs. In a fascist state, by contrast, the demands of the nation as an abstract entity are given precedence over human requirements. The state's imaginary existence is perceived as being far more real than reality.

In the absence of a larger, more humane context, of a living connection between individuals, based on the acknowledgement of their mutual needs, the state can come to play a broader role than naturally warranted, masquerading as father, mother, "big brother" and god all rolled into one. Yet, as it was never intended to perform these roles, it can never hope to play them well.

Democracy, and its legitimate surrogate a republic, despite any short-term deficiencies, ultimately works, and works well, because it is naturally integrative. It forms a reliable basis for stability and progress by effectively including everyone. The greatest threat to the democratic process, apart from the prospect of our individually failing to prosper, or of our collectively failing to progress, is that there are people who simply do not believe in it—or who would prefer not to believe in it—as it does not suit their ends. Ironically, like the Biblical Pharisees, who are the first to say "Lord, Lord" (Matthew 7.21), those who think less of freedom are typically the ones to boast more about it, cloaking themselves beneficently in democracy's collective rites and rituals, while acting deliberately against it by condemning as traitorous, unpatriotic, or outright inimical, the views of those who would define their freedom differently. Such superficial and outspokenly verbal proponents of "freedom" are, at the same time as they are cheering for the nation, not typically the greatest believers in mankind's true liberty, progress, dignity, or potential. In their cold and sterile "realism" can be found distilled only selfish fear in the present, and a dark and stagnant resignation to the inevitability of future slaughter.

It is typical of human behavior, for people to look for ways to distinguish themselves and to brand themselves as superior. Once they identify a convincing basis for that presumed superiority, they then, in the manner of the emperor's new clothes, deliberately group together with those whose comparable opinions will uniformly reinforce their own. The resulting prideful assuredness is neatly accomplished through "groupthink" without impacting reality directly.

A teacher might, for example, see his opinion as superior to a layman's, due to the fact that he is more educated. A pastor might think his own opinion more moral. A businessman, in regarding both views as trending toward the impractical, might accentuate the value of his own opinion as being far more "down to earth." And a farmer might well discount all such views as having an insufficient connection with life, with having been formulated by people with "no dirt under their fingernails." Here the value of democracy may be identified as the inverse of our individual limitations, while at the same time founded on an acknowledgement of our individual worth, for where we can all agree from the perspective of our separate viewpoints, chances are we may be on to something. While the decisions thus arrived at may be more contentious and belabored, they are also likely to have the stabilizing advantage of being more acceptable to all, and hence more sufficient, and enduring. Yet democracy itself will endure only if we are consciously mindful, and actively supportive of its conditions.

Should we come, in the modern age, to finally acknowledge the validity—if not, in the end, the value—of our widely-divergent views, they would remain part of a broader problem, for even while being uniquely-articulated, they should be as intelligently-framed. For democracy will not work where the voters vote unwittingly against themselves. The farmer should know what legislation proposed may affect the marketing of his crops, the businessman the success of his business, the priest the honoring of his values, and the educator, his ability to educate his students. To the extent that each does not know, the whole of society cannot appropriately respond. This subtle and corrosive form of ignorance emerges, not so much or so often from any culpable lack of understanding, as from the absence of a more basic awareness.

We will not move to address the problems we cannot see. The midnight decrees of shadow governments, deals forged behind closed doors which, while guided by narrow interests, extend outward to impact society as a whole, decisions rendered secretly by those who presume to always know best, are like fungal growths flourishing in dark places that call for the light of awareness. This makes knowledge, education, and a genuinely free press, persistent demands of a democratic society, for they are the perennial adversaries of oppression.

While there has been a notable decline in malignant nationalism, among developed nations, in recent years, this can be attributed, less to an improvement in moral outlook, than to a shift in competitive emphasis. With the emergence of a global marketplace, the competitive strivings of nations have been given an economic twist, thus mitigating many of the more extreme and outwardly violent forms of nationalism. While this is indeed a favorable development, the mistake should never be made of regarding nationalism as simply a reflex of economics, or as just another bourgeois myth. Nor are commercial values refined enough, in themselves, to advance the concept of nationhood beyond the level of savagery.

A nation reflects in the coarseness or refinement, the benignity or malignity of its organically-developed traditions and laws, the quality of life of its people. Malignant nationalism is not a self-generating force, but a by-product of deprivation, meaninglessness, harshness, and backwardness. Such are the root causes of malignant nationalism, and of the violence which attends it. These more basic conditions must be addressed if the potentially catastrophic impact of malignant nationalism, in the modern age, is to be blunted.

The transition to an integrative world is not a threat to nationalism per se, but merely to its violent excesses. Divested of any destructive overtones, manifesting through acts of war, in favor of our prosperous and inclusive participation in an increasingly integrative world, nationalism becomes something which we are, for the first time in history, more in a position to appreciate than to fear. The role of the nation itself is being progressively transformed by modern trends to serve the collective good internationally, adding to rather than subtracting from the strength of our ever-more integrative efforts.

Conclusions

Confronted with two radically divergent futures, reflecting the wonders or terrors of science, we find ourselves now at the crossroads. The choice of a positive future is one we must make through innovative thought and compassionate action, not merely through bland or hypocritical statements of preference or hope for what is good, for it is the most crucial choice which mankind, as a species, will ever be called upon to make.

Through technology's influence, we are beginning to manifest at the material level, what Hindus and Buddhists claim has always existed at the root spiritual level—the fundamental unity of life. As knowledge itself becomes more integrative, its growth in the scientific sphere is compelling us to advance in the social sphere beyond the level of isolated interests, material values, and bigoted views. The science underlying nuclear weaponry represents a knowledge of processes penetrating beyond the ordinary limits of Newtonian science into the quantum realm beyond the atom, demanding a corresponding advance in our collective wisdom. Whether we will be up to the challenge that this transition portends, remains an open question, upon whose outcome our future survival will be determined.

In the film *Force 10 from Navarone*, demolitions expert, Miller, instructs two members of his infiltration team to detonate an explosive charge designed to collapse a dam in enemy-occupied territory. The dam's destruction is, in turn, to release a raging wall of water sufficient to topple a bridge further down the river, a bridge over which a substantial enemy force is to be caught advancing. An explosion occurs, yet the dam remains standing. Miller's comrades are beside themselves with anger and disappointment, but Miller himself remains calm. "Let Nature take her course," he advises. Within minutes the first visible breaks in the dam appear. They continue to increase in size and number. Shortly thereafter, the entire structure collapses. The flood waters are released and the bridge downstream is destroyed.

At the beginning of the 1980s, the Soviet Union, as a military superpower, appeared to be on the rise and unassailable. The Russian war in Afghanistan raged against the humiliating backdrop of our failed Iranian rescue mission, and amidst the nagging perception that the United States had already been surpassed militarily by its chief international rival. Yet, even then, at the moment of such pyrrhic triumphs and apparent strategic advantages, the Soviet system was poised for an inevitable collapse; the cracks in its overall structure were becoming increasingly noticeable and wide. Nature was taking her course. Within a decade, and as climaxed in 1991 by the failed right wing coup in Moscow, Soviet Communism (though not, alas, Russian authoritarianism) came to an abrupt end. It was an end that no one ten years before would have easily predicted. Such an outcome occurred without a single shot being fired in open warfare against The Soviet Union by an adversarial power.

At present, we may despair of ever seeing an end to traditional warfare, as we had once despaired of seeing an end to Soviet Communism. Yet Nature, here as well, is taking her course, and we, as a part of Nature, are coming along for the ride. The global movement away from traditional warfare is proceeding apace amidst changes already substantially underway in the social, economic, personal, religious, and cultural spheres of human activity. War, particularly in its nuclear variant, is becoming increasingly untenable, unnatural, unmanageable, and manifestly counterproductive to most—if not all—concerned. Where it is not, is typically where some have been unnaturally left behind, itself a contradiction of progress which modern conditions prompt us to comprehensively address.

At much the same time, the competitive drive that would seem to be an inherent part of our evolutionary makeup is being refined, elevated and favorably morphed. It is, to an ever-increasing degree, being redirected to higher and more productive realms of alternative activity. Such partial transformations are but part and parcel of the larger continuing process of global integration.

Doing a rewind on history, it is hard for us to see peace as natural and war as unnatural, where the historical record overwhelmingly suggests the opposite. A key difficulty at present is that we are burdened by the past, by the perspective of where we have come from, while not yet aware of precisely where we are going. This makes the present transitional phase particularly problematic.

But Nature *will* take her course; her prevailing trend is to sweep us, with the force of a favorable tide, away from traditional warfare. In the process, and through the momentum already unleashed with the splitting of the atom, she will either draw us in with the current of unprecedented scientific advances and the riches of peaceful development, or crush us with the tidal force of insecurity, security paranoia, and increasingly destructive and risk-laden conflicts. The question remains, whether we will pick up the hint with alacrity, and adopt the elevated norms associated with our dramatically changed environment. Or will we punish ourselves needlessly yet again, by recklessly engaging in what purports to be a unprecedentedly catastrophic war, on terms as yet unknown, with this unsatisfactory outcome proceeding as the seemingly sole and unequivocal means of reminding ourselves of what we should, in our stubborn ignorance and, by now, already understand—that to secure the future, the violence of the past must be abandoned.

Our epic rivalry with The Soviet Union, whether more as a result of our unrelenting opposition or of that system's own internal collapse, has been finally and permanently resolved. Along with it has fled the greatest immediate threat to planetary life. Yet other indicators that we are drifting toward a broader ruin remain.

1. Our knowledge continues to precariously outpace our wisdom, as we continue to exhibit no notable tendency to systematically redress that perennial imbalance. As our solution to the problem of deadly viruses created by the use of strong disinfectants has been to douse the germs with even stronger disinfectants, our inelegant and inadequate solution to the problem of weapons and war has been to develop increasingly more devastating weapons that enable increasingly catastrophic wars. Our front line response to terrorism has likewise been—and for the most part remains—nothing more creative or thoughtful than to apply terror's destructive equivalent back against its perpetrators, in a way that will less offend our presumably more refined sensibilities. Until we discover a more satisfactory, wise, and inclusive range of responses, we will remain locked within the confines of a primitive mentality and more violent plane of action, from which there will be no conceivable end to putting ourselves perpetually in harm's way.

2. Driven by a rabid consumerism, we remain distracted from the meaningful substance of life, and from our living connection to one another, by a kaleidoscopic array of fragmentary pleasures and advantages that we feel we must—yet can never really—secure. For their sake, and without deeper consideration, we repeatedly engage in mindless heated conflict with one another. Being killed in a stampede at Walmart on Black Friday has become almost a symbol, and not just a symptom, of our debased cultural level. When our center is untrue, so too are we. Not true to ourselves (as our souls) we cannot be happy, and from the womb of our unrelieved dissatisfaction emerges the warlike specter of our strife. Needless conflict, excessive competition, and a mean sense of scarcity, have become the consistent hallmarks of our age, casting us even further adrift from where we now—more than ever—need to be. Most troubling of all is how we weave the discontent of others into the very fabric of our individual happiness, operating under the

false and hazardous premise that we exist apart from others in a world that is, in truth, increasingly one. We thus make firm the assurance that violence, terror and war, will never come to end until (as a consequence of their technologically-magnified impact) we do.

3. Our society, as arguably the most influential worldwide, continues to send mixed messages about violence. From the old six-gun Westerns, to more contemporary action films, to video games that are currently all the rage, we have consistently celebrated both guns and the figure of the rogue gunman. Yet we are invariably shocked and bewildered whenever an immature youth or undiscriminating adult, too damaged or confused to separate fiction from life, casts himself in the ever-iconic role. When the predictable outcome is tragic, our reaction is always the same: "How could this have happened?"—though we should know enough originally not to ask. Yet before we find time to properly pose this question, before the bodies are buried, or the living have been apportioned their suitable space to grieve, a ravenous press characteristically leaps in like clockwork to bestow upon the murderers the vicarious immortality that had inspired the latter's violence to begin with. This standard, predictable and lemming-like role of "the news," as an unwitting enabler of violence, neatly paves the way for the next set of terrorists or shooters seeking a comparable publicity. As there were evils that primitive tribes were too terrified to speak of, for fear of invoking them, perhaps the identities of these modern-day demons—though, by no means, their atrocities or the lessons we must learn from them—should not be spoken of directly, let alone bandied about promiscuously by the media until their names are as familiar to us as those of our cars or our relatives.

4. We remain, as a whole, a very odd species. We refuse to accept, arguing the point to the brink of mass-annihilation, what we can never really hope to change, i.e., that the future, however similar to the present or the past, will be in its own essential nature different. All of our former glories will and must remain just that, and any more spectacular future that we may now conceivably envision, will be no less transitory at the time it is eventually realized than the seemingly less-satisfactory period we are impatiently living through now; as the hand of time ticks forward, and change relentlessly occurs, we must remake ourselves and our world anew, sculpting raw change into a more gratifying level of progress.

At the opposite extreme of our delirium, we refuse to change what we can never really accept, i.e. that there may be, for us and for our progeny, no future at all, no options of precarious result to choose from; we may instead see only the tragic and comprehensive end to our promising human beginning. As we march drone-like into the field of our workaday activities, we think that nothing will change where, in fact, everything—at least everything material—does. In this altered technological world, where quantum truths have struck a note of Heisenbergian uncertainty that resonates throughout all that we do, and throughout all that we once came to view with an ironclad assurance, we still prattle distractedly about what is practical, without truly understanding what's real. It is indeed a very different world now, one in which our outdated Clausewitzian "business as usual" can only end predictably with the sunset of humanity in the dawn of the final war. Even sooner than that, we would inevitably bear witness, should we simply stay on course, to the erosion, in the name of security, of our cherished liberties, to the loss of that which makes life itself inherently worthwhile, for the makeshift solutions we come up with at the eleventh hour of any extreme crisis can hardly be as satisfactory as the results of a due foresight, duly applied now. We nonetheless continue to defend our routine ignorance as if it were a precious jewel. Too few of us have the temperament to proactively consider all views, let alone to empathize with our enemies in their plight, as John Kennedy did—and needed to do—during the Cuban Missile Crisis. Should the stakes be as high again, will our leadership be as worthy or as prepared?

Now, no less than before, the nations of the world cannot and should not be expected to, in any way, abandon their need to pragmatically defend themselves. What we now need to beneficially create and foster is an upgraded mode of international environment in which they won't need to resort to arms, or will do so far less often. It is what we establish in the present when things are quiet that will determine what we must face in the future, when they are not. In the midst of any attempt to avert the ultimate war, it is essential for us to acknowledge, as Shakespeare wrote, that "the fault lies not in our stars but in ourselves." We are not doomed, in spite of our noblest efforts, but only by the persistent absence of any genuine effort at all. We have shown up perfunctorily and repeatedly on lurid battlefields of tragic devastation; we should be as willing to fight, as reliably and as tenaciously, for the wiser and more enduring ends of morality and truth. War and peace, as these pages have suggested, begin and end where they have always proceeded, within each of us initially, in our hearts, minds and continuously evolving natures.

From the standpoint of a very realistic optimism, looking ahead to our spiritual future rather backward to our biological past reveals Nature to be, not an enemy, but a friend in our most hopeful and peaceful collective aspirations. Its impact and course subtly underlie our more limited human actions as a positive and uplifting force, one that would encourage us in our evolutionary advance, and whose own natural outcome is not one aligned with mass human conflict—most certainly not at this stage of our epic journey through time—but with the overarching imperatives of a broader planetary future.

The choice as to what future awaits us, enlightened or ruinous, is ours alone to make, and the time for that choice is now. As to that choice's ultimate outcome, be it wondrous or tragic, and yet to be determined, as expressed by Paramahansa Yogananda:

> Though the works of the human race disappear tracelessly by time or bomb, the sun does not falter in its course; the stars keep their invariable vigil. Cosmic law cannot be stayed or changed, and man would do well to put himself in harmony with it. If the cosmos is against might, if the sun wars not in the heavens but retires at due time to give the stars their little sway, what avails our mailed fist? Shall any peace come of it? Not cruelty but goodwill upholds the universal sinews; a humanity at peace will know the endless fruits of victory, sweeter to the taste than any nurtured on the soil of blood (291).

Bibliography

Aristophanes. *Lysistrata. Project Gutenberg.* Web.

Arnold, Edwin. *The Light of Asia.* London: Library of Alexandria, 1879. *BuddhaNet.* Web.

Babbitt, Irving, trans. *The Dhammapada.* New York: Oxford UP, 1965. Print.

Bach, Richard. *Jonathan Livingston Seagull.* New York: Avon, 1973. Print.

Bartlett's Familiar Quotations Revised and Enlarged. Boston: Little, Brown and Co, 1980. Print.

Bassford, Christopher. *The Clausewitz Homepage.* 27 Feb. 2013. Web.

Bergen, Peter. *Holy War Inc.: Inside the Secret World of Osama bin Laden.* New York: Free P, 2001.

Blavatsky H.P., ed., *The Voice of the Silence: Chosen Fragments from the "Book of the Golden Precepts."* Pasadena: Theosophical UP, 1976. Print.

"Blind Men and the Elephant." *WordFocus.* 2013. Web.

Bradley, Omar N. *A General's Life: An Autobiography By General Of the Army Omar N. Bradley and Clay Blair.* New York: Simon and Schuster, 1983. Print.

Brodie, Bernard. *The Absolute Weapon: Atomic Power and World Order.* New York: Harcourt, Brace, 1946. Print.

Buchanan, Scott, ed. and Benjamin Jowett, trans. *The Portable Plato.* New York: The Viking Press, 1948.

Carrera, Jaganath. *Inside The Yoga Sutras: A Comprehensive Sourcebook for the Study and Practice of Pantanjali's Yoga Sutras.* Buckingham, VA: Integral Yoga Publications, 2005. Print.

Carruth, Gorton. *American Facts and Dates.* New York: Harper & Row, 1987. Print.

Casablanca. Dir. Michael Curtiz. Perf. Humphrey Bogart, Ingrid Bergman, and Paul Henreid. Warner Bros, 1943. Film.

Clarfield, Gerald H., and William Wiecek. *Nuclear America.* New York: Harper & Row, 1984. Print.

Columbia University Faculty, eds. *Introduction to Contemporary Civilization in the West, Vol. II* New York: Columbia UP, 1948. Print.

Colossus: The Forbin Project. Dir. Joseph Sargent. Perf. Eric Braeden, Susan Clark, and Gordon Pinsent. Universal, 1970. Film.

Condon, Richard. *The Manchurian Candidate.* New York: Four Walls Eight Windows, 1959. Print.

Darwin, Charles. "The Origin of Species." *Introduction to Contemporary Civilization in the West.* Vol. II. New York: Columbia UP, 1948. Print.

Dooley, Mike. *Choose Them Wisely: Thoughts Become Things!* New York: Simon & Schuster, 2009. Print.

Dyer, Wayne W. *The Power of Intention: Learning to Co-create Your World Your Way.* London: Hay House, 2004. Print.

Easwaran, Eknath. *Essence of the Bhagavad Gita: A Contemporary Guide to Yoga, Meditation & Indian Philosophy.* Tomales, CA: The Blue Mountain Center of Meditation, 2011. Print.

Easwaran, Eknath, trans. *The Upanishads.* Tomales, CA: The Blue Mountain Center of Meditation, 2007. Print.

"Edmund Burke." *Respectfully Quoted: A Dictionary of Quotations.* New York: Barnes and Noble, 1989. *Bartleby.* Web.

Einstein, Albert and Leopold Infeld. *The Evolution of Physics.* New York: Simon & Schuster, 1938. Print.

Evans-Wentz, W.Y. *Tibetan Yoga and Secret Doctrines.* New York: Oxford UP, 1967. Print.

Faulkner, Harold Underwood. *American Economic History.* 6th ed. New York: Harper, 1949, Print.

Fitzgerald, Frances. *Fire in the Lake: The Vietnamese and the Americans in Vietnam.* Boston: Little, Brown & Co, 1972. Print.

Force 10 from Navarone. Dir. Guy Hamilton. Perf. Robert Shaw, Harrison Ford, and Edward Fox. Columbia, 1978. Film.

Freud, Sigmund. *Civilization and its Discontents.* Trans and ed. James Strachey. New York: W.W. Norton, 1961. Print.

Geldern, James. "1934: Socialist Realism." *Soviet History.* 2013. Web.

Goldstein, Joseph. *The Experience of Insight: A Natural Unfolding.* Santa Cruz: Unity P, 1976. Print.

Hammer, Joshua. "Risking Friendly Fire." *Newsweek* 4 Mar. 1991. Print.

Hartmann, R.R.K. and F. C. Stark, F.C. *Dictionary of Language and Linguistics.* New York: Halsted Press, 1972. Print.

The Holy Bible: Catholic Action Edition. Gastonia: Good Will Publications, 1953. Print.

Herken, Gregg. *Counsels of War.* New York: Knopf, 1985. Print.

Hicks, Esther and Jerry Hicks. *The Vortex: Where the Law of Attraction Assembles All Cooperative Relationships.* London: Hay House, 2009. Print.

Hitchens, Christopher. "Thomas Jefferson: The Pirate Wars: To the Shore of Tripoli." *Time Magazine.* 5 July 2004. Web.

Hobbes, Thomas. *Leviathan.* Baltimore: Penguin, 1976. Print.

Hogg, Ian V. and J. B. King. *German and Allied Secret Weapons of World War II.* London: Phoebus, 1976. Print.

Humphreys, Christmas. *Zen: A Way of Life.* Boston: Little, Brown, & Co, 1962. Print.

Hughes, H. Stuart. *Contemporary Europe: A History.* Englewood Cliffs: Prentice-Hall, 1976. Print.

Katzer, Julius, ed. and trans. *Lenin's Collected Works.* Vol. 31. Russia: Progress, 1966.

Kaplan, Fred. *The Wizards of Armageddon.* New York: Touchstone, 1984. Print.

Kennedy, Jacqueline. *Historic Conversations on Life with John F. Kennedy: Interviews with Arthur M. Schlesinger, Jr., 1964.* New York: Hyperion, 2011. Print.

Kirsch, Adam. "Heralding the End of War." *The New York Sun.* Web. 2 Jan. 2008.

Knierim, Thomas. "Uncertainty Principle." *Big View.* April 2013. Web.

Lama, Dalia and Howard C. Cutler. *The Art of Happiness: A Handbook for Living.* New York: Riverhead Books, 2009. Print.

Lewis, Bernard. *The Crisis of Islam: Holy War and Unholy Terror.* New York: Random House, 2003. Print.

Lipton, Bruce H. and Steve Bhaerman. *Spontaneous Evolution: Our Positive Future (And A Way To Get There From Here).* New Delhi: Hay House, 2009. Print.

Maeterlink, Maurice and Alfred Sutro. *The Life of the White Ant.* London: George Allen & Unwin, 1927. Print.

Massachusetts Institute of Technology Faculty, eds. *The Nuclear Almanac: Confronting the Atom in Peace and War.* Reading: Addison-Wesley, 1984. Print.

McNeill, William H. *History Handbook of Western Civilization.* Chicago: U of Chicago P, 1953. Print.

Mill, John Stuart. *Utilitarianism. Project Gutenberg.* Web.

Milton, John. "On His Blindness." *Bartleby.* Web.

Milton, John. *Paradise Lost. Project Guttenberg.* Web.

Morris, Richard ed. *The Encyclopedia of American History.* 6th ed. New York: Harper & Row, 1982. Print.

Mosley, Leonard. *On Borrowed Time: How World War II Began.* New York: Random House, 1969. Print.

New International Bible. Bible Gateway. Web.

Newhouse, John. *War and Peace in the Nuclear Age.* New York: Knopf, 1989. Print.

Ni, Hua-Ching, trans. *I Ching: The Book of Changes and the Unchanging Truth.* Los Angeles: Seven Star Communications Group, 2007. Print.

Orwell, George. *1984: A Novel.* New York: New American Library, 1983. Print.

Pascal, Blaise. *Pascal's Pensèes.* New York: E. P. Dutton & Co. *Project Guttenberg.* Web.

Peers, Allison, ed. and trans. *St. John of The Cross, The Dark Night of the Soul.* New York: Doubleday, 1959. Print.

Plotinus. *The Enneads.* Trans. Stephen McKenna. London: Faber & Faber, 1956. Print.

Poe, Edgar Allan. *The Masque of the Red Death.* Ed. Edward H. Davidson. *Selected Writings of Edgar Allan Poe.* Boston: Houghton Mifflin, 1956. Print.

Prem, Sri Krishna. *The Yoga of the Bhagavat Gita.* Dorset: Element, 1988. Print.

Press Association. "Stephen Hawking: Mankind must colonise space or die out." *The Guardian.* 9 Aug. 2010. Web.

Prury, Nevill. *Dictionary of Mysticism and the Occult.* San Francisco: Harper & Row, 1985. Print.

Rinpoche, Kangyur. *The Treasury of Precious Qualities: A Commentary on the Root Text of Jigme Lingpa.* Boston: Shambhala Publications, 2010. Print.

Ritter, Allan and Julia Conway Bondanella, eds. *Rousseau's Political Writings.* New York: W.W. Norton & Co, 1988. Print.

Rosa, Joseph G. and Robin May. *An Illustrated History of Guns and Small Arms.* Secaucus: Castle Books, 1974. Print.

Rubinstein, Alvin Z. "The Soviet Union and the Peace Process Since Camp David." *Soviet Foreign Policy in a Changing World.* Eds. Robbin Fredrick Laird and Erik P. Loffmann. New York, Aldine, 1986. 774-778. Print.

Saraswati, Swami Satyanada. *A Systematic Course in the Ancient Tantric Techniques of Yoga and Kriya.* New Delhi: Yoga Publications Trust, 1981.

Shapiro, Walter. "Cold War Remnant: Cancer for Baby Boomers." *Politics Daily.* Web. 2010.

Shirer, William L. *20th Century Journey: A Memoir of a Life and The Times, Vol. II: The Nightmare Years, 1930-40.* Boston: Little, Brown & Co, 1984.

Smith, Huston. *The Religions of Man.* New York: Harper & Row, 1965. Print.

Stearns, Raymond Phineas, ed. *Pageant of Europe: Sources and Selections from the Renaissance to the Present Day.* New York: Harcourt Brace & World, 1961. Print.

Stevens, William Oliver and Allen Westcott. *A History of Seapower.* New York: Doubleday, 1942. Print.

Stoessinger, John. *The Might of Nations.* New York: Random House, 1973. Print.

The Terminator. Dir. James Cameron. Perf. Arnold Schwarzenegger, Michael Biehn, and Linda Hamilton. Hemdale, 1984. Film.

The 13th Warrior. Dir. John McTiernan. Perf. Antonio Banderas, Vladimir Kulich, and Dennis Storhøi. Touchstone, 1999. Film.

Toffler, Alvin. *Future Shock.* New York: Random House, 1970. Print.

Tolkien, J.J.R. *Lord of the Rings Trilogy.* London: George Allen & Unwin, 1955. Print.

Tsipis, Kosta. *Arsenal: Understanding Weapons in the Nuclear Age.* New York: Simon & Schuster, 1983. Print.

Tuchman, Barbara. *The Guns of August.* New York: Random House, 1962.

Tzu, Lao. *Tao Te Ching.* Trans. R.B. Blakney. New York: Signet, 2007. Print.

Webster's New Explorer Desk Encyclopedia. Springfield, MA: Merriam Webster, 2003. Print.

Whyte, William. *The Organization Man.* New York: Doubleday, 1957. Print.

Williams, Samuel Cole. *Adair's History of the American Indians.* New York, Promontory P, 1930. Print.

Wilmott, Ned and John Pimlott. *Strategy and Tactics of War*. London: Marshall Cavendish Books Limited, 1983. Print.

"Words of Wisdom from the Buddha." *Beliefnet*. Web.

Yogananda, Paramahansa. *Autobiography of a Yogi*. Los Angeles: Self-Realization Fellowship, 1994. Print.

Yutang, Lin. *The Wisdom of India and China*. New York: Random House, 1943. Print.

Zukav, Gary and Linda Francis. *The Heart of the Soul Emotional Awareness*. New York: Simon & Schuster, 2001. Print.

Index